Roller Coasters, Rockets & Computers

plus the odd space elevator

Ian East

www.openchannelpublishing.com

Roller Coasters, Rockets & Computers, plus the odd space elevator

Published by Open Channel Publishing Ltd. in 2016.

ISBN : 978-0-9565409-9-7 (premier edition, paperback, colour)

www.openchannelpublishing.com

To Nicola

for my redemption

To Arthur and George

for joining me on many roller coasters

Cover images

- launch of Space Shuttle Discovery at the Kennedy Space Center, 31st May 2008, on mission STS-124 to deliver and install two Japanese modules on the International Space Station
- SeaWorld Orlando's *Manta* roller coaster in May 2011
- Inmos T9000 Transputer modules, interconnected via a C104 Virtual Channel Processor.

Content

About the author

Ian East began his career as an Engineering Physicist in the defence industry before returning to university to gain a Ph.D. in physics from Imperial College, London, in 1983. He then continued as a University Lecturer in physics, and later computing, in diverse institutions over the next twenty-five years, publishing two textbooks and around thirty papers. He has also served as a consultant and contractor to commercial and military institutions.

Ian now spends his time writing, publishing and implementing his own programming language, Honeysuckle, when not flying gliders or home-built radio-controlled planes.

He lives with his wife, two sons and a large sleepy Labrador near Oxford.

Preface

As a youngster, I was turned on by technology and science. I was turned *off* by school science. (There was no school technology.) Like just about everything else we were 'taught', it seemed irrelevant and far from inspiring. Luckily, I had already found inspiration in the fabulous machines I saw at the Farnborough Air Show, including an airliner that flew like a fighter (Concorde) and the similarly unique Harrier, that could take off and land vertically. I had also heard about radar and electronic navigation from my Dad — a technician in the RAF, during the war. Then there was the "space race" and science fiction, which then offered a positive view of the future. *Thunderbirds* and early *Dr. Who* were a delight on TV, but books by Arthur C. Clarke changed my life, especially *Against the Fall of Night* and *The Lion of Commarre*, which told of machines that replicated and maintained themselves, removing any need for human labour. The idea that life could be so much better stayed with me.

I simply could not understand how school science could be so boring.

Only when I began research for my Ph.D. did I finally discover just how wonderful science and technology could be. "Sit down, shut up and listen" was finally over. The words of one of the school staff, Donald Banner, returned loud and clear: "there is no such thing as teaching, only learning". I had to take control and figure things out myself, asking for help only when I really needed it.

Over the years, I have been repeatedly shocked at the "dumbing down" of science in schools. Worst of all, mathematics has been all but eliminated from physics. Physics is the beating heart of science. Chemistry is merely a branch of physics, as Dirac pointed out in the 1920s. Biology is at last slowly coming into the fold as well — a revolution in our time. Until now, it has remained purely empirical. What's different about physics is that it is *model-driven*. A model (theory) is predictive and thus acts as a font for hypotheses, which can be tested against observation. Deviations from the predicted outcome of experiment or observation are then used to improve or revise the model, which must therefore be precise — *i.e.* mathematical. Mathematics is the language of 'formal' science.

Removing maths from physics is like eliminating speech from Shakespeare.

Mathematics without object and context, as is normal in school, exposes a different problem. Mathematicians are a breed apart. While they are quite happy starting out with pure abstraction, the rest of us prefer to arrive there through *generalization*. We need to see similarity and pattern in application — in the way we resolve a number of distinct problems.

To put the maths back into the physics aids the learning of both. We can repeatedly see how the former is applied, and we can properly appreciate the power of the formal model in the latter.

While arguably the most critical benefit from learning science is to understand something of the world around you, a very close second is an appreciation of how we can affect that world. This runs through technology and beyond to encompass the consequences of putting it to use. We are fortunate to enjoy a degree of democracy, which confers a responsibility to educate ourselves. To be *rational* means to adapt our perception to the facts; to adapt the facts to our perception is to be irrational. Science and mathematics convey a value placed upon care with both evidence and inference.

Science offers not only our best chance for prosperity, it allows our only chance of survival.

If we allow our young to grow up believing whatever they like, without reliable evidence or rational interpretation then either we will not survive or we will wish we hadn't. There are far too many "flat Earthers" who believe in racial superiority or that burning fossil fuel can have nothing to do with global warming. As H. G. Wells once said, "civilization is a race between education and catastrophe".

At the same time, it is a mistake to expect too much from science. It cannot explain "life, the universe and everything", only one thing in terms of another. The idea that it fundamentally conflicts with religion is simply wrong. It merely conflicts with *fundamentalist* religion, which is a very different thing. Arguably, it can also remove conflict *between* religions.

In so doing, science might bring peace as well as prosperity.

This book is written with home education in mind, though its use is open to any tutor. The study method I use is just the oldest one on Earth, based around the 'tutorial'. I first go over and illustrate the day's material, then leave my son to read for himself and do some exercises to apply the new ideas. The cycle ends with a second tutorial to verify he has correctly understood what he has done. While this may sound formal, it is, in fact, social, and thus a pleasure. It works just as well, if not better, with a group of up to six or so. As every tutor knows, numbers matter, and there are definite limits to group size.

I have not attempted to describe how, for example, a model roller coaster could be built. I am not convinced this would have too much value, though it might well be fun. There is nothing wrong with imagination, which needs exercise like everything else. The idea that learning should always be fun is one I reject. Most of the time, you just have to roll up your sleeves and do the work.

You have to commit yourself, as President Kennedy said, to do the things that are hard.

My hope is that the book can also prove useful to those responsible for larger classes, and in a similar way. A segment can be reproduced on the whiteboard and the ideas applied as class- or homework.

In my humble opinion, it remains vital that students also read the book for themselves.

No attempt has been made to address any syllabus in its entirety. That was never the objective.

There follows a summary of the topics covered, by chapter and section.

Technical content

Chapter 1 *Roller coasters*

Generic and transformable nature of energy; potential and kinetic energy; work and energy; Galileo's Law of gravity; conservation of energy; Newton's Laws; energy of body in a gravitational field; MKS units and the Système Internationale (SI); commutation (in multiplication); radian measure; energy conversion (transformation); rolling resistance; air resistance; resolution of vectors; mechanical equilibrium; sine/cosine/tangent; small angle approximation; Newton's Second Law in terms of momentum; 'δ' notation "for small change in"; circular motion; centripetal acceleration and force; 'g' as unit of acceleration; Newton's Law of Gravitation; planetary/geosynchronous orbit; effect of camber; physiological limits to acceleration; relativity of motion; negative energy; chemical and nuclear energy; constancy of velocity of light; mass as energy; conservation of mass/energy.

Chapter 2 *Planetary climate*

Energy *vs* power; electric, magnetic & electromagnetic (EM) fields; thermal energy; temperature as 'hotness'; hotness *vs* heat; absolute (Kelvin) temperature scale; thermal radiation; Stefan-Boltzmann Law; radiative equilibrium; reflection/absorption/emission of EM radiation; specular reflection; planetary albedo; absorption/emission by black body; solar power output, incidence on Earth; luminosity and inverse-square law; solar constant; solar energy capture; atomic & molecular absorption/emission of EM radiation; simple & compound greenhouse effect; optical depth; atmospheric thermal transport (evaporation/condensation); paleoclimatology; feedback, positive & negative; science *vs* religion.

Chapter 3 *Space*

Velocity required for low & geosynchronous Earth orbit (LEO, GEO); Earth equatorial velocity; sidereal day; escape velocity; Tsiolkovsky Rocket Equation; exhaust velocity of chemical & nuclear rocket; Δv as measure of propellent; payload ratio and step rocket; consequences of Rocket Equation for interplanetary return (must refuel at destination); tower, in compression & tension; specific strength/length.

Appendix: incommensurability & irrational numbers ; Euler's number *e*; differentiation, of polynomials and logarithms; integration & (first) fundamental theorem of calculus.

Chapter 4 *Computers*

Information and memory; logarithmic measure of information; signal vs noise; sampling & integration; quantization ('digitization'); digital *vs* analogue signal/record; register, state & value; register base (radix) *vs* cost; representation of (signed) number; synchronous *vs* asynchronous communication; synchronization; system compositionality/modularity; propositional calculus & Boolean algebra; logic gates; compositional logic; state machines (automata); fabrication of gates from switches; latch & flip-flop; register transfer; arithmetic logic with "look ahead" carry; machine instructions; fetch/execute cycle; conditional branch; stack and load/store architecture; assembly language; instruction execution pipeline; time/space diagram; memory requirements; memory hierarchy & caching; mean cost & delay; memory-mapped i/o; processor interruption; parallel computation; social impact of (unbounded) automation.

Acknowledgements

I would like to express my thanks to Barry Cook Ph.D. C.Eng. for taking time to review the work and make many very helpful comments and suggestions. Any goofs that remain are down to me, not him.

Thanks also go to my son Arthur for his useful observations on Chapter 4.

Chapter 1

Roller coasters

1.1 Finding the energy, and keeping it

Do you have the potential?

Energy comes in many guises: it's the heat in your morning coffee; the reason why a car is difficult to stop; what's locked up in the gasoline that fuels it; and the work you do in pedalling a bike up a hill. It's all about movement: the movement of atoms and molecules, which constitutes heat; the movement of your car along a road; or the movement of you up that hill.

The one thing we know for sure about energy is that it is *never* created or destroyed. It's only ever transformed from one kind into another. Even mass is just energy. If you could mop up all the material blasted out by a nuclear explosion, you'd find just a little bit missing: the minute amount which was converted into all that heat, light and concussion.

A roller coaster car needs *kinetic* energy — the energy of motion. One way to get that energy is to drop the car from a high point, letting it roll down hill. The question is what height do you need to go as fast as you'd like? Before we can answer that, you must first decide on a speed.

Step 1 Decide how fast you'd like your roller coaster car to go, and express it in metres per second. (A mile is ~1600 metres.)

You'll find out in later steps whether you've chosen too great a speed. (Either it will kill the riders, when they go around a curve, or the track will need too much space.)

You might think that we'll need to know the weight, or rather the *mass*, of the car.
You'd be wrong. At least, for now.

We need to start from four principles :

Energy represents the capacity to perform work.

Work = force applied × distance moved

Force = mass × acceleration (Newton's Second Law)

Gravity imparts a constant acceleration, independent of mass. (Galileo)

The energy 'lost' when a mass m falls from a height h is therefore given by :

$$E_g = m \times g \times h$$

where 'g' denotes the acceleration due to gravity. (We shall henceforth simply write 'mgh'.)

Exactly the same energy is 'consumed' (work is performed) — an amount mgh — when we move the same mass up to that height.

Now, we assert a fifth principle :

Energy is *conserved* — it is never lost, but only converted from one form into another.

Although work has been performed, energy hasn't been consumed at all. Instead, it's been converted into something we call *potential* energy. There must be energy there, even if the mass isn't moving, since there is still the capacity to perform the same amount of work, on the way down.

Of course, to push the mass up to height h, we have to find energy mgh from some source. If we use a winch, powered by an electric motor, plugged into the main supply, we will get an electricity bill. That bill will count the number of 'kilowatt-hours" we 'consumed' (converted from electrical energy to gravitation potential energy).

It's time to talk about units.

Clearly, we need some sort of standard whenever we measure something. For example, here we've chosen to measure distance, or height, in metres. To calculate energy, we have to multiply things together, which are each measured in their own units. A speed is measured by counting metres travelled in some measure of time. If we choose a second as the standard unit of time then we can measure speed in "metres per second", or 'm/s'. Acceleration measures speed gained per unit time, and so "metres per second per second", or 'm/s^2'. For example, the acceleration due to gravity is approximately (to a single "significant digit") ten metres per second per second, so we write :

$$g \simeq 10 \text{ m/s}^2$$

To calculate force, we multiply acceleration by mass. (Multiplication always gives the same result, whichever way around you do it — $(m \times g) = (g \times m)$. It is thus said to 'commute'.) So, we must decide on a unit of mass to count. We shall choose the 'kilogramme' — a thousand grammes.

This system of units is called MKS — metres, kilogrammes, and seconds — or the *Système Internationale d'unités* — SI, for short. In these terms, the unit of force is called a 'newton'[1] and is that required to accelerate a mass of one kilogramme at the rate of one metre per second per second :

$$1\,\text{N} \equiv 1\,\text{kg} \times 1\,\text{m/s}^2 \equiv 1\,\text{kg m/s}^2$$

To quantify mechanical energy, or work done, we multiply force applied by distance travelled. A unit of energy is named after another British physicist, James Joule :

$$1\,\text{J} \equiv 1\,\text{N} \times 1\,\text{m} \equiv 1\,\text{N m}$$
$$\equiv 1\,\text{kg m/s}^2 \times 1\,\text{m} \equiv 1\,\text{kg m}^2/\text{s}^2$$

Notice how you can (with mechanical systems) always reduce the units to metres, kilogrammes and seconds (MKS). The *Système Internationale* merely extends MKS with names for more abstract quantities, honouring the scientists concerned, and adds a typographic convention.[2]

It's common to confuse energy with power. Power is energy converted per unit time. Hence, one 'watt' is defined as one joule per second :

$$1\,\text{W} \equiv 1\,\text{J/s} \equiv 1\,\text{kg m}^2\text{s}^{-3}$$

Note that unit notation is often rendered a little more convenient by writing a denominator as a negative power. Another convenience is the use of the prefix 'k' for 'kilo', 'M' for 'mega', 'G' for 'giga' *etc.*, which denote ten to the power three, six, nine and so on. The measure by which your electricity company assesses your consumption can easily be converted to joules :

$$1\,\text{kWh} \equiv 1{,}000 \times 3{,}600\,\text{J}$$
$$\equiv 3.6 \times 10^6\,\text{J}$$
$$\equiv 3.6\,\text{MJ}$$

Now we can put some numbers in and calculate the energy we might acquire from a drop :

— height h say, 60m

— mass m say, 4,000 kgs.

---

1 According to SI convention, we deny Isaac Newton a capital letter when referring to a unit of force, *except* in single letter abbreviation. So we write 1 newton, or 1 N. We do the same with units of energy and power — joule (J) and watt (W).

2 In fact, four more fundamental physical phenomena must be quantified and receive a metric under the SI: temperature (kelvin), electrical current (ampere), luminosity (candela) and matter (mole — a measure of the number of atoms).

The mass, remember, must account for a typical load of passengers — say, eight rows of four — plus the carriage itself, which will weigh quite a bit, to carry so many people. A typical adult human weighs sixty to a hundred kilogrammes. An ordinary car now weighs a thousand.

$$\begin{aligned} E_g &= m\,g\,h \\ &\simeq 4{,}000 \times 10 \times 60 \text{ joules} \\ &\simeq 2.4 \text{ MJ} \end{aligned}$$

Notice that this is just two thirds of a kilowatt-hour — the equivalent of the energy used when leaving a small heater on for just forty minutes. We might have expected more, especially when we consider how hot the brakes might get, were we to use them to draw to a halt.

But then how fast are we going?

Any moving mass has the capacity to perform work. Should a train collide with a vehicle on a level crossing, it will push it along some way, until friction brings it to a halt. It thus has energy.

We call it 'kinetic' energy, and shall denote it by E_k.

Since we'd like to keep our example simple, and ignore things like friction and air resistance for now, it's handy to think instead of a scenario out in space. Suppose we'd like to move a satellite, of mass m, using a rocket. If the rocket applies a constant force then we should expect a constant acceleration, leaving the satellite with some final velocity v, after a time t.

Any measure of velocity is purely *relative*. You have no option but to assess your speed with respect to some "frame of reference". Essentially, this means you can choose whatever scale you like, as long as you are consistent. With a satellite, it makes most sense to measure your speed with respect to the surface of the planet directly below. Let's suppose, on that scale, we started with zero velocity.[3]

Recall that force equals mass times acceleration. Acceleration equals change in velocity divided by time. If, after time t, the final velocity is v then the acceleration is v/t. Distance travelled is given by average velocity times time. Since acceleration is 'uniform' (constant), and we started at rest, the average velocity is just half the final value, $v/2$.

This would mean that the energy the satellite acquires is :

$$\begin{aligned} E_k &= \textit{force} \times \textit{distance} \\ &= m \times \frac{v}{t} \times \frac{v}{2} \times t \\ \therefore \quad E_k &= \tfrac{1}{2}\, m v^2 \end{aligned}$$

Now we can calculate the speed of our coaster after the drop.

3 This would mean that the satellite is in rather a special orbit, referred to as 'geostationary'.

If all the energy from the drop is converted into kinetic energy then we can equate the expression for it with that for gravitational potential energy.

$$E_k = E_g$$

$$\therefore \quad \tfrac{1}{2} m v^2 = m g h$$

$$\therefore \quad v = \sqrt{2 g h}$$

Notice, first and foremost, that it doesn't seem to matter how heavy the carriage is. In fact, this is just what Galileo Galilei observed back in the 17th Century.

Notice, second, that the units on each side agree. (Taking the square root of m^2/s^2 yields *m/s*.)

If we plug in the height we proposed (60m), we find we should expect a velocity of 24.5 m/s, or just under 90 kph (kilometres per hour) — about 55 mph (miles per hour).

This assumes that no energy is wasted in heat, vibration or noise. In practice, there will be some losses, but not enough to change things significantly, if care is taken in the design of our system : the wheels must be smooth and roll freely, and the track must be both smooth and rigid.

To see how high we need the drop, given the speed we'd like, we must invert the above equation, bringing h to the left-hand side :

$$h = \frac{v^2}{2g}$$

Application

1.1 90 kph is not really fast enough to provide the thrill we might want. Decide a speed you think would prove sufficiently exciting then calculate the drop height h required. (Remember to convert any value so it is ultimately measured using metres, kilogrammes and/or seconds.)

1.2 What must we do to the drop height in order to raise speed by ~50% (multiply it by 1.5) ?

1.3 Assuming carriages, as described above, are hoisted to the top, ready for the drop, by an electric winch, at the rate of one every 2 minutes, every day of the week, calculate the effect on the monthly electricity bill. (Assume a charge of 15p per kWh.)

1.4 What happens to speed and energy 'consumption' when we add a second identical carriage to form a train (assuming we retain the same drop height) ?

Keep on rolling

We would not expect a carriage rolling along a track to keep on going forever, at the same velocity, but *why not*? Just what is going on to slow it down?

To see what happens between wheel and rail, we would need a really good microscope… or that more powerful weapon, a good imagination. The surface of neither rail nor wheel will be perfectly smooth — to make it so would probably be technically impossible. Even if we knew how to make a perfect crystalline surface, there would still be bumps formed by individual atoms. Given traditional machining, the surfaces we actually use will look like a moonscape, with sufficient magnification. There will be hills, valleys, ravines and mountains.

If each hill on our wheel were to exactly match each valley on our track, there would be no resistance to its rolling, but this is hardly likely. We might instead imagine the worst that can happen: a big ugly mountain on our wheel coming up against a big ugly (though different) mountain on our track.

It will never be worth analysing the precise details of each track and each wheel to determine the exact forces involved, which would have to account for every combination of wheel orientation and track location. As is often the case with real engineering, we need a simple model which can account for the variation in easily measured parameters, such as the materials employed and vehicle weight.

When you buy a car which weighs twice what your old one did, you should not be surprised that more power is needed to push it along. A little experimentation in the lab reveals that the force that acts against the car — the "rolling resistance" — varies in direct proportion with weight. To be precise, it varies with the *load* on each wheel concerned. If there are four wheels bearing the car, the load on each will be just a quarter of the car's weight.

To overcome rolling resistance, and roll at a constant velocity, an equal and opposite force is required (Figure 1.1). This may seem obvious, but it took a genius to spell it out:

A body continues at constant velocity until acted upon by a (net) force. (Newton's First Law)

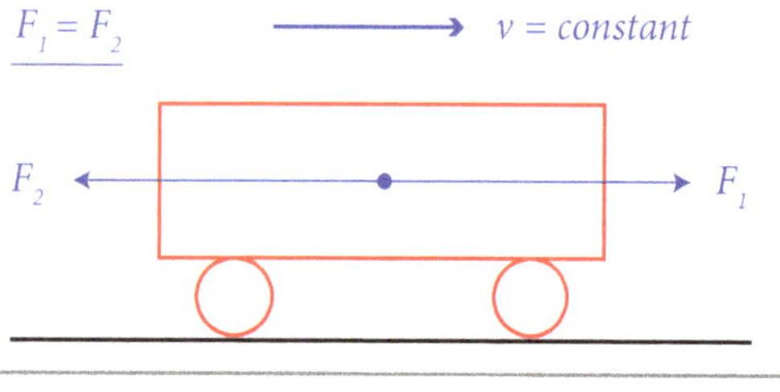

Figure 1.1 A carriage at constant velocity.

In other words, a 'force' is whatever it is that causes *acceleration*. Newton was simply doing what any true scientist does: defining his terms before using them. (His Second Law defined 'mass', which measures that property of a body which determines how much acceleration you see when applying a given force.) If you want a constant velocity (no acceleration) then forces must add up to zero.

Rolling resistance also depends upon the materials used and how they are prepared. A rubber tyre on a tarmac road will offer at least *four times* the resistance as an iron wheel on an iron rail. Because we will need four times the force to propel the vehicle at a constant velocity, we will be converting four times the energy per metre travelled. Road transport is vastly less efficient than rail.

We can capture the effect of the particular wheel and track chosen in the constant of proportionality between rolling resistance and load:

$$F_{rr} = C_{rr} \times F_g$$

The value of C_{rr} may change, as the surface of both wheel and track wears.

Our roller coaster carriage has no engine. Rolling resistance will cause it to slow down unless we do something. It turns out that rolling resistance is equivalent to facing an up-hill gradient. So, one thing we can do is to provide an equal and opposite gradient *down* hill.

Figure 1.2 depicts the geometry of our carriage rolling down hill. To understand the diagram, we need to first ask *why we don't fall through the floor.* This brings us back to Newton:

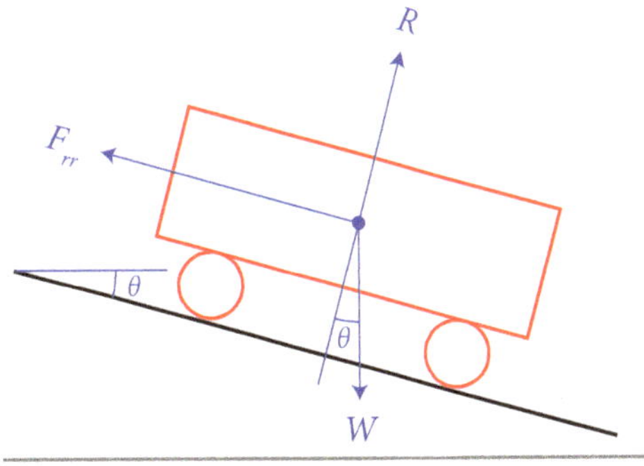

Figure 1.2 Carriage on down-hill gradient.

In equilibrium, every force is met with an equal and opposite reaction. (Newton's Third Law)

'Equilibrium' means no acceleration, in any direction; which, in turn, means no net force. To prevent our carriage falling through, the track is somehow counteracting its weight. But how? It does not appear to be doing anything.

Again, we would need a good microscope to see what is going on. In fact, just like the floor on which we stand, the track acts like a spring, *deflecting* a tiny amount. (It's a very powerful spring, and so does not need to bend much to counter a huge load, which is why we don't usually notice the change.) The reaction R is just its "restoring force".

Next, we need to understand that we can resolve any force in two directions, at 90°. We can then form the triangle of forces in Figure 1.3. A little trigonometry leaves:

$$F_{rr} = W.\sin\theta$$
$$F_g = W.\cos\theta$$

By dividing the first equation by the second, we get:

$$F_{rr} = F_g.\tan\theta$$

If we compare this with the earlier expression for F_{rr}, we can see that:

$$\tan\theta = C_{rr}$$

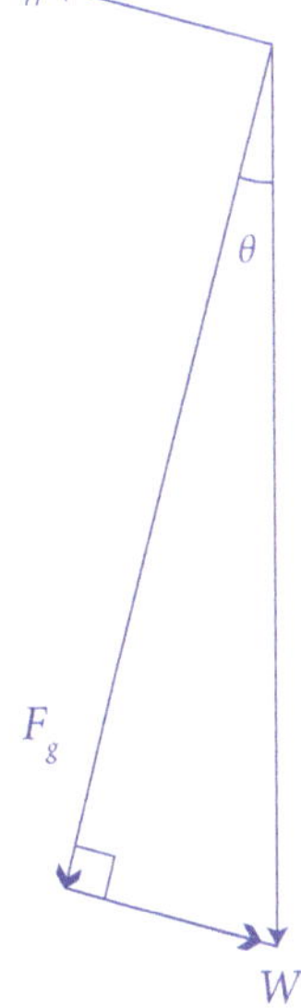

Figure 1.3 Resolving vectors

The gradient required equates with the coefficient of rolling resistance. But how do we put this into effect, given that we want the track to rise and fall anyway — perhaps even include a vertical loop?

All we need to do is to ensure that the track always descends the right amount on each stretch. Anyone who has seen a real roller coaster may have noticed that a vertical loop terminates below its entry. The same happens with a hill — the descent finishes below the start of the initial ascent — and the reverse with a dip — the ascent finishes below the start of the intial descent.

If we know the length L of the complete track, following all the hills, dips, curves and loops, we can work out how far below the start the finish must be, before brakes, and perhaps a final ascent, bring the carriage to a halt:

$$h_{rr} = L.\tan\theta = L \times C_{rr}$$

That's all we need to do, to counter rolling resistance.

Application

1.5 Calculate the height h_{rr} required to counter rolling resistance when the track length L is 3,000m and the coefficient of rolling resistance C_{rr} is 0.002 (steel wheel on steel rail).

1.6 Now compute the height required should the wheels have rubber tyres, rolling on a smooth concrete track, where C_{rr} is 0.010.

1.7 If nothing is done to counter rolling resistance, the carriage would eventually come to a halt. Assuming the carriage has steel wheels and rolls on a straight and level steel track, with a starting velocity of 100kph, how far would it get?

Why does neither speed nor mass affect the deceleration?

Where has all the initial (kinetic) energy gone?

Hair resistance

Rolling resistance is not the only force slowing down a carriage.

As you zoom along, your hair will not behave well, and you will feel another force acting against your face. Air may be invisible, but it has a way of letting you know it's there. What's happening is that you, your fellow riders and the carriage are pushing the air out of the way. To calculate the force applied, and hence the reaction, we need a more precise statement of Newton's Second Law:

Force applied equals rate of change in momentum.

'Momentum' is the tendency of a body to continue moving, or a measure of the action needed to reduce its movement or bring it to a halt. It can be reduced to two factors: mass × velocity. Since mass is typically constant, the Second Law is often quoted simplistically as $F = m\,a$, where a denotes the rate of change in the velocity, *i.e.* the acceleration. Here, it is different; we must think of our carriage as a piston, forcing air out of the way, at a constant velocity. It is *mass* that is changing — the mass of air in front of the piston. Force can then be calculated as the mass of air moved per second δm_a multiplied by its velocity, which is equal to that of the piston v.[4]

The volume of air moved per second is given by the area A of the piston (our carriage) multiplied by the distance it travels, which, in one second, is just v metres. (Remember, velocity is distance per unit time, *i.e.* metres per second.) Density measures mass per unit volume, so the mass of that air is given by its density ρ_a multiplied by $A \times v$:

$$F_{ar} = \delta m_a v = (\rho_a A v)v = \rho_a A v^2$$

Note that the effective value of A will be a bit less than the full cross-section of carriage and passengers facing along the track. The factor by which this is reduced is called the "drag coefficient" c_d and depends upon how 'streamlined' things are — how easily air can slip by.

Clearly, unlike rolling resistance, *air* resistance is not independent of velocity; in fact, it increases as the *square* of velocity. It's useful to see what velocity is required for it to equal rolling resistance:

$$\rho_a A v^2 = m_c g C_{rr}$$

$$v = \sqrt{\frac{m_c g C_{rr}}{\rho_a A}} = \sqrt{\frac{4000 \times 9.8 \times 0.002}{1.2 \times 2.0}} \simeq 5.7\text{m/s}$$

This is the kind of "back of the envelope" calculation often carried out by scientists and engineers. Since the calculation can be done as precisely as you like, by calculator or computer, the time saved is in using a "rule of thumb" value for each parameter. For example, the density of air varies

4 To make variables stand out from ordinary text, physicists often use the greek alphabet. The letter 'δ' has special significance. It doesn't denote a variable itself, but is read as "a small change in" whatever the following symbol represents. All mathematics can be read out as plain English (or any other natural language). If in doubt, always ask someone how.

according to altitude, temperature and humidity, but 1.2 remains correct to two significant digits, which is all we need.

The result tells us that air resistance exceeds rolling resistance beyond ~6m/s. Given that it increases as the *square* of velocity, it will dominate our consideration. (We decided on a speed roughly four times as high, which means a force due to air resistance *sixteen* times that due to rolling resistance.)

Because our intention is for the carriage to have a constant velocity, we might counter air resistance with an additional incline θ_{ar}, leading to the need for a further elevation of the track by h_{ar}:

$$mg\sin\theta_{ar} = \rho_a A v^2$$

$$\therefore \quad \sin\theta_{ar} = \frac{\rho_a A v^2}{mg}$$

When an angle is small, its tangent approximates its sine (because the adjacent side has length almost equal to that of the hypotenuse), so:

$$h_{ar} = L\tan\theta_{ar} \simeq \frac{L\rho_a A v^2}{mg}$$

That pesky square causes a problem. For the kind of numbers we've chosen thus far, we'll find that an additional elevation of around 100m would be required — not very realistic! Clearly, everything possible must be done to reduce the effective area A. One possibility is to lengthen the carriage, or form a train, which increases m at the same time. While this will help, it is unlikely to be enough.

There would seem no choice but to break the track into two (or more) sections and winch the carriage up for another drop, if we're to rely on gravity alone.

Application

1.8 To reduce air resistance, the track is divided into two sections, each of length 1,500m, a second identical carriage is added to form a train, with combined mass 8,000 kgs, and streamlining improved to limit the effective area A to just 1 m^2. Calculate the additional elevation h_{ar} required to fully counter air resistance and maintain a velocity of 20m/s.

1.9 Where does all the extra energy go?

1.10 How do hills and valleys affect air resistance and the amount of counter-action required?

1.2 Going around in circles

Use the force!

It wouldn't be much of a roller coaster if it simply took you along in a straight line. Some would say it wasn't much good without an inversion or two (or six or more). Apart from adding to the illusion of "going somewhere", twists and turns add the thrill of *acceleration*, just like the drop.

Whether it's a horizontal bend, a vertical loop, or a "barrel roll" (following a corkscrew path, as though around and along the inside of a barrel), it's acceleration that adds the buzz.

Combat pilots call acceleration (inaccurately) a "g force". One reason why the fighter planes of the future will likely have no pilots is that they may lose consciousness if they experience much beyond about "10 *g*" (ten times the acceleration due to gravity) for more than a few seconds. (The figure varies with both individual and experience, and is open to dispute.) Suffice it to say there is a limit, and people can't fight — or fly — too well when they're out cold.

Acceleration can be 'positive' or 'negative' — taken to mean upward or downward, respectively. Pilots experience both, but are usually happier being forced into their seat than out of it —we are, after all, better upholstered to resist the seat than the shoulder straps — but both are thrilling. It's the same turns that can produce either effect, depending on whether our heads point outward or inward.

Not many roller coaster riders will be quite as fit as a fighter pilot, so ought to stay well below 10 *g*. The Space Shuttle was constrained to a maximum of around 3 *g*, on lift off and re-entry, so there's no need to subject folks to much more, given a typical crew description of their ride. A 'dragster' manages only around 4 *g*, never mind a Ferrari. It may surprise some to learn that a mere glider can offer up to 7 *g*, during aerobatics, providing it's built to that purpose. A drop is pretty exciting, and that's just −1 *g* (note the minus sign), though perhaps the sensation of 'falling' adds quite a bit.

So how do we figure out how sharp to make the curve, to get the buzz we're after?

Figure 1.4 shows a mass moving around a circular path — perhaps a spacecraft orbiting the Earth. At any instant, we can describe its motion via a 'vector' representing its velocity. Any vector can be drawn as an arrow, whose length depicts its magnitude, and whose orientation shows its direction. While the magnitude of the velocity v of an orbiting spacecraft may not change, its direction certainly does, and continuously.

Therefore, it *must* be accelerating.

We can work out the acceleration with a little geometry.

After a little time, say δt, the craft has moved through an angle $\delta\theta$. Figure 1.5 shows how things have changed.

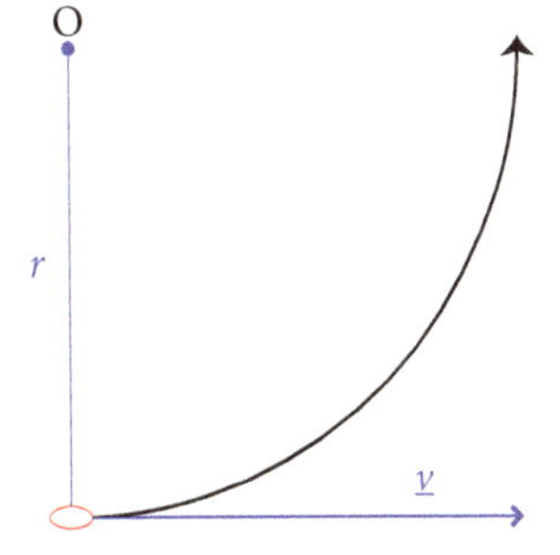

Figure 1.4 Circular motion.

Things are easier if we measure an angle in *radians*. We just divide the length δx of the arc described by that of the corresponding circle's radius r:

$$\delta\theta = \frac{\delta x}{r}$$

If the craft is moving with a constant 'angular velocity' ω radians per second, we can write:

$$\omega = \frac{\delta\theta}{\delta t} = \frac{1}{r}\frac{\delta x}{\delta t} = \frac{v}{r}$$

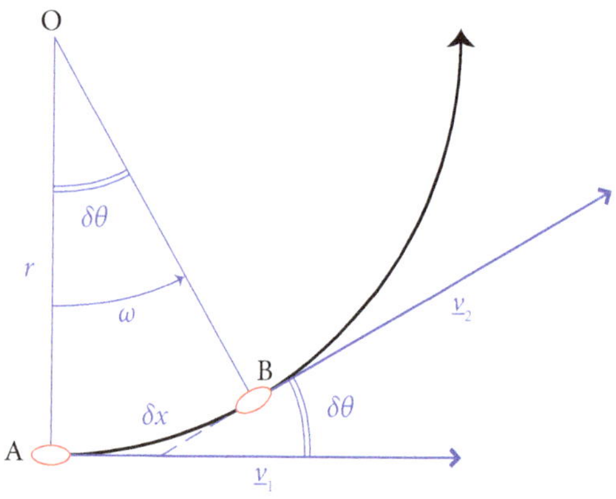

Figure 1.5 Geometry of circular motion.

Note that v is just the magnitude of the linear velocity. As a vector, $\underline{v}$ also has *direction*, which is constantly changing, as shown in Figure 1.5. Both magnitude and direction of change $\underline{\delta v}$ that occurs in time δt, *i.e.* the *acceleration* a, can be found from the triangle of vectors, shown in Figure 1.6.

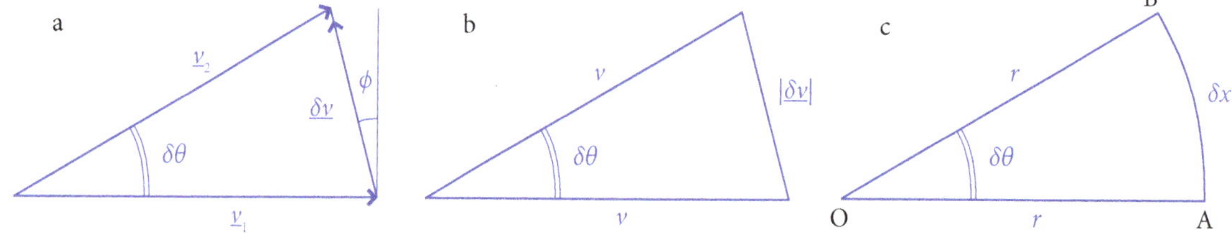

Figure 1.6 The same triangle, formed by velocity vector, velocity magnitude and arc.

By applying our earlier definition of angle (in radians):

$$\delta\theta = \frac{|\delta v|}{v}$$

The acceleration of any orbiting object depends solely on its velocity, linear or angular:

$$a = \frac{|\delta v|}{\delta t} = v\frac{\delta\theta}{\delta t} = \frac{v^2}{r} = r\omega^2$$

For a body to remain in a circular orbit, *something* must provide a centripetal[5] force F_c:

$$F_c = ma = \frac{mv^2}{r} = mr\omega^2$$

5 You may have heard of something called a 'centrifugal' force that acts outwards, away from the centre. There is no such thing. The idea arises from a misunderstanding of circular motion.

We have yet to account for the *direction* in which this acceleration must apply. What is the angle φ between $\underline{\delta v}$ and the normal to $\underline{v}_1$?

Figure 1.6a forms an equilateral triangle, so the remaining angles are equal and sum to $\pi - \delta\theta$, which leaves the angle between $\underline{\delta v}$ and the normal:

$$\varphi = \frac{\pi}{2} - \left(\frac{\pi - \delta\theta}{2}\right) = \frac{\delta\theta}{2} \qquad (\pi \equiv 180^{\circ})$$

(Because π is defined as the ratio of a circle's perimeter to its diameter, it also measures a half-circle (180°) in radians — half the perimeter divided by half the diameter.)

If the net effect of uniform circular motion through $\delta\theta$ is an acceleration directed along a line at half that angle then we can conclude that acceleration is always toward the centre of rotation.

For a satellite in orbit around a planet, the centripetal force will be provided by gravity. If the planet has mass M, Newton's Law of Gravitation determines the velocity needed for an orbital radius r:

$$F_c = \frac{mv^2}{r} = F_g = \frac{GMm}{r^2}$$

$$\therefore \quad v = \sqrt{\frac{GM}{r}}$$

Note that the mass m of the satellite is immaterial.

G is one of those parameters which appears to be arbitrary, and not a consequence of anything else, as far as we know. Like the velocity of light c, it just accounts for the way the universe is made. We believe it to be the same everywhere within the universe, and at every point in time. But then we haven't even left our solar system yet, and we cannot explain all we see in the sky.

Unfortunately, our roller coaster is unlikely to be able to take advantage of a passing planet. Instead, we must rely on the track to push our carriage around.

Raiding the bank

One way of supplying the necessary centripetal force is to add a camber (bank) to the track, as shown in Figure 1.7. The reaction R to the weight of the carriage can be resolved into two components:

$$R.\sin\theta_c = \frac{mv^2}{r}$$

$$R.\cos\theta_c = mg$$

We can divide the first equation by the second to reveal the necessary camber to turn the carriage around an arc of radius r at velocity v:

$$\tan\theta_c = \frac{v^2}{rg}$$

Note again that the mass of the carriage is immaterial. Only the velocity and radius matter.

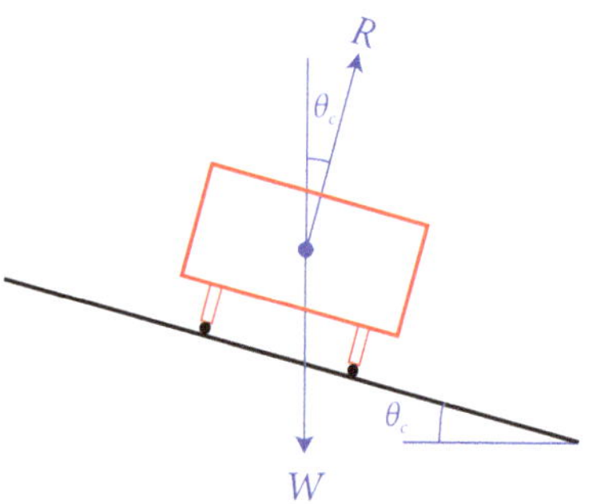

Figure 1.7 Carriage on cambered track.

Application

1.11 What camber (bank) is required for a velocity of about 100kph and a curve of radius 100m? Use sensible approximations.

1.12 What acceleration will the riders experience, measured with regard to g (the acceleration due to Earth gravity at sea level)? Give an answer rounded to the nearest integer.

Best not kill the rider

We often hear talk of 'g-force', with regard to fighter aircraft and roller coasters. As with "centrifugal force", there is no such thing. All there is is plain old *acceleration*.

What causes the confusion is that it is common to measure acceleration in relation to that imposed (mysteriously) by gravity. Since gravity is something with which we are all familiar, it is only sensible to use g as a unit. Hence, you might hear a fighter pilot speak of experiencing '5g' in a turn.

There are very good reasons why only the young and the fit are recruited to train as fighter pilots. Acceleration upwards imposes a greater burden on the heart: it must then apply a greater force on the mass of blood which must reach the brain each second. When the body accelerates downward, it must similarly work harder to pump that same blood *out* of the brain each second.

Safety requires a lower bound on downward acceleration because of the risk that a blood vessel in the brain may burst if the pressure rises too much, causing irreparable damage or death. The corresponding problem with upward acceleration is less dangerous, directly. If the brain receives *insufficient* blood for just a few seconds, the rider/pilot just loses consciousness. (Of course, it will be equally dangerous if you're busy fighting for your life at altitude, sitting astride a jet engine.)

Fighter pilots refer to these two situations as "eyes in" or "eyes out".

Lateral (sideways) acceleration carries a different threat: it can break your neck.

The remaining variety is arguably under-exploited in the design of roller coasters. We can tolerate high acceleration *forward*, which is why a racing car can be very exciting. Even a 1 g leap from the grid can deliver a thrill. However, the truth is that an internal combustion engine is very poor at delivering acceleration from a standing start. It delivers very little power when it begins to turn. It's "power curve" drops off sharply from a peak, typically at several *thousand* revolutions per minute (RPM). While it makes an exciting noise, revving up and dropping the clutch has a limited effect.

Electric motors, while silent, are far better, as they remain powerful at "low revs".

The problem for those planning a roller coaster is that high forward acceleration is difficult and expensive to achieve. A catapult of some kind is required, as with an aircraft carrier or glider launch. In comparison, the 1 g experienced in a drop is much easier and cheaper to deliver. (One of my more popular rôles as a Dad was as a high place from which my little boys could fall onto the sofa.)

There is also comfort to consider. While a forward acceleration delivers a thrill, in a properly designed seat with a headrest, backward acceleration feels like crashing, and so will be rather less popular. Lateral acceleration is much the same.

The duration of the acceleration can make a big difference to safety. Pilots can survive seven or eight g for ten or even twenty seconds; ordinary mortals should normally stay within six, for no more than ten seconds. But these are just rules of thumb. We really are all different, in so many ways.

We all have to assess risk and balance it with pleasure. Some ride; others prefer the telly.

Application

1.13 A car's acceleration is typically quoted via the time (in seconds) required to reach 60mph from a standing start. Calculate the "nought to sixty" that corresponds with 1 g.

Looping the loop

Entering a vertical loop, gravity will slow the carriage as it rises; we are again exchanging kinetic energy for gravitational potential energy. Unlike the horizontal case, the velocity v is not uniform, even when the radius of curvature r is constant. From the conservation of energy, we learn something interesting. We know that the carriage will rise (or fall) a height equal to twice the radius, r so:

$$\frac{1}{2}mv_{base}^2 - \frac{1}{2}mv_{top}^2 = mg2r$$

$$\Rightarrow \qquad \frac{v_{base}^2}{r} - \frac{v_{top}^2}{r} = 4g$$

Centripetal acceleration at the base of a circular loop will *always* be exactly $4g$ greater than that at the top, no matter what values we choose for v_{base} and r. Inward acceleration at the top must never fall below $1\,g$ or the carriage will fall from the rails under gravity. The situation becomes worse when we remember that riders at the bottom already endure $1g$, due to gravity, before we push them up. Upward acceleration at the base will thus be at least $6\,g$, which is too high to be either comfortable or safe for most people. If a downward loop were executed, it would be even more unhealthy.

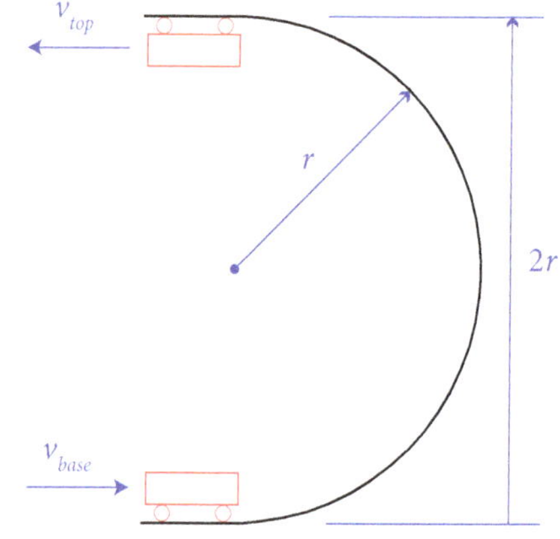

Figure 1.8 A vertical half-loop.

We need to establish two distinct inequalities:

$$\frac{v_{top}^2}{r} \geq 1g \qquad\qquad \frac{v_{base}^2}{r} \leq ng$$

where n is the g limit applied to a healthy human (assuming unhealthy ones do not ride).

We might design a loop with centripetal acceleration at the top close to (but not less than) $1\,g$, which will give a *perception* of "zero g", as experienced by astronauts in orbit.

We're used to being tugged and *not* falling.

From the above, we can obtain a joint inequality:

$$\frac{v_{base}^2}{r} - \frac{v_{top}^2}{r} \leq (n-1)g$$

Substituting the earlier relation, leaves us with what we have already determined verbally, that the smallest vallue of n would be five, giving a total acceleration experenced at the base of $6g$.

The only answer is to vary the radius of curvature.

If we increase the radius of the lower quadrant (quarter loop), both on the rise and the fall, the centripetal acceleration can be reduced to something more tolerable — say $4g$, giving a total of $5g$, which is acceptable for the few seconds required, and will add to the thrill of the ride. However, we must also take care not to increase the height of the loop, which would add to the energy lost and so to the speed needed on entry, and the consequent centripetal acceleration. A *decrease* in the radius of the upper quadrant can compensate for the increase in that of the lower one, leaving the height unchanged, and can also raise the centripetal acceleration at the top. This further reduces the velocity, and thus acceleration, needed at the base.

The result is a "teardrop shape", familiar to riders of modern coasters.

The first equation above still holds, subject to:

$$r_{base} + r_{top} = 2r$$

We can capture the change via a parameter α:

$$r_{base} = \alpha r \qquad 1 < \alpha < 2$$

$$r_{top} = (2 - \alpha) r$$

The second equation is revised, taking the equalities only:

$$\frac{v_{base}^2}{r} - \frac{v_{top}^2}{r} = \left[(n+1)\alpha - 2\right] g$$

Combining both equations, we get:

$$(n+1)\alpha - 2 = 4$$

$$\Rightarrow \qquad \alpha = \frac{6}{n+1}$$

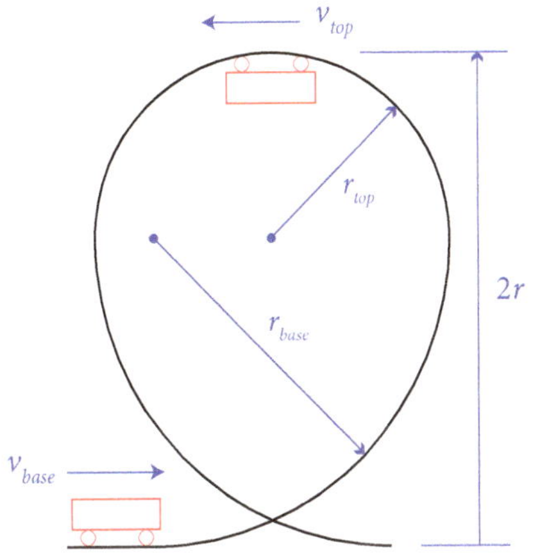

Figure 1.9 A vertical tear-drop.

So, for example, if we choose a base centripetal acceleration of $4g$, we need to multiply the base radius by 1.2, giving a ratio of 3 : 2 between that and the top radius. The total experienced would then be $5g$, allowing for gravity's contribution, which should be acceptable for the few seconds required.

Application

1.14 Making sensible assumptions for all parameters, design yourself a life-size roller coaster, complete with drops, turns and loops.

Remember, a vertical loop can be upward, with the carriage upright, or downward, with the carriage inverted. It might even twist half way. A half upward loop might precede a drop, and so on.

1.15 Now design a toy one, perhaps with a marble as the vehicle.

How do the drops, turns and loops scale with vehicle mass?

You might try building this one, improvising for the track and support.

1.3 So you think you understand energy

Probably, the earliest idea around was that of 'work'. If the lord of the manor wanted a hill moving, or a forest cleared, best make yourself scarce. A hill or forest twice the size would take twice the effort. Close behind the notion of work, followed that of 'power'. If he who must be obeyed had fallen off his high horse into a ditch, it would take twice the muscle if he was twice as fat.

It would have to wait for machines to arrive before anyone became aware of energy. A steam engine of twice the power would consume twice the coal per day. And coal was expensive stuff.

Science tagged along *behind* invention, playing "catch up". It wasn't long before people began wondering just what it was in the coal that mattered, and where it went — you couldn't burn ashes. After reviving the ancient Greek idea that everything is made from indivisible little balls called 'atoms', they eventually figured out just how heat and work were related. What united them was energy. Heat is just the aggregate energy of many atoms whizzing, or wobbling, around each other. The early steam engines merely converted heat into mechanical power, which is just energy per unit time.

We've seen how energy can be stored up by lifting a mass to the top of a hill, and how such gravitational *potential* energy can be converted back to the tangible kinetic kind. But gravitation can also do the opposite. Imagine a rogue planet, drifting between the stars. From its own perspective, it has no energy. From our perspective, when it wonders into our solar system, it might have tremendous kinetic energy. Because velocity is relative to the observer, kinetic energy is so too. Provided the velocity relative to the sun is not too great, the planet might end up in orbit around it. Let us suppose that it was at rest with regard to the sun when it first began to feel the tug of gravity. The point is that, rather than easily release energy, we would have to *put energy in* to restore the previous state of affairs.

Since the free planet began with no energy, its potential energy while in orbit around the sun must have been *negative*. The closer to the sun it fell, the deeper the "potential well".

The same applies to electrons around the nucleus of an atom. All that happens in a chemical reaction is that an electron is lifted out of one potential well and dropped into a another. If the second is deeper than the first, energy will be released, typically as heat. Some 'activation' energy has to be invested to liberate the electron from its initial home, but in an 'exothermic' reaction there is still some left over after that is repaid. Essentially, this is what happens when you burn coal.

You release *chemical* energy.

Nuclear energy is actually simpler. Until Ernest Rutherford discovered what happens when you fire positively-charged particles at them, atoms were thought to be like plum puddings, with their ingredients mixed together uniformly. What he saw could only be explained if all the protons were packed tightly in the centre, rather like a football at the kick-off point in a huge stadium. All those like-charged particles had to be held together by a force far more powerful than the electric one associated with chemical energy. The right missile — one with no charge — could cause havoc.

As indeed it did, with such tragic consequences, over poor Nagasaki and Hiroshima.

One of the grandest truths discovered is that energy is never destroyed when it's used to perform work. It just turns up in other forms, in the same amount. Energy is *conserved.* This is stranger than you might think. There must be a given total in the universe if it is finite, as we now believe it is. Why so much and not more, or less? Our interpretation of various astronomical observations is that the universe had a beginning. Where did all that energy come from?

Science may yet be able to answer these questions. What it can never answer is *why* the universe exists at all. A void can break no rules, and would seem the more likely thing.

The final word here must go to Albert Einstein, who discovered a still greater truth.

All that had been learned about the strange electric and magnetic forces had finally been captured in four beautiful equations, brought together by James Clerk Maxwell, but there was a problem.[6] The laws of physics should remain the same wherever you measure from. If you measure from the platform or a moving railway carriage, your calculations may look different but the rules should not vary. Newton's Laws were proof against a change to the "frame of reference". Maxwell's equations were not.

Others, notably the legendary Hendrik Antoon Lorentz, had tried to resolve this issue and failed. It took Einstein to establish his "principle of special relativity", that a theory remains unaltered by the frame of reference, as paramount, and determine that other assumptions were wrong.

Both time and length ceased to be absolute. *They* could change with the frame of reference.

Two more truths emerged: nothing can overtake light, and $E = mc^2$.

The significance of the second discovery was never lost on physicists, but it took a while to sink in with the rest of us. Any mass is equivalent to a gigantic amount of energy — c is a very big number; c^2 is *enormous.* We now know that it can be released by colliding a particle with its 'antiparticle'.

But the significance goes much deeper than the discovery of an all-but-infinite resource. We have learned that mass is just another manifestation of energy.

Energy, though seen in so many forms, is all there is.

6 Credit for the beauty of the equations, as we now know them, should go to Oliver Heaviside, who reformulated each using the vector calculus he himself had developed.

Chapter 2

Planetary climate

2.1 Sun worship

One thing we're pretty sure of now is that there is exactly the same amount of energy in our universe now as at the time of its creation. Energy is never either created or destroyed. All we can ever do is convert it from one form to another, according to our needs. An electric heater serves as an example, converting electrical energy into heat. We measure the effectiveness of a heater by its *power*, and the cost of using it in terms of the *energy* it gets through. A 1kW heater is usually enough to keep us warm in a typical room on an average day. The two are of course intimately related. Power is just the rate at which energy is converted. Leave the heater on for an hour and you add a unit (kWh) to your bill.[1]

A typical power station converts the chemical energy in, say, natural gas into electricity with an output of perhaps a gigawatt (GW, a million kW). Current global human power consumption, of all kinds, amounts to about 12.5 terawatts (TW, a thousand GW). The output of the Sun is around 385 *thousand billion* terawatts. Stars maintain this for billions of years.

We might respect nature and be grateful.

In 1820, Hans Christian Oersted and André-Marie Ampère discovered that an electric current produces a magnetic field and can thus cause motion. Eleven years later, Michael Faraday demonstrated the reverse — that motion through a magnetic field could induce an electric current. By 1865, James Clerk Maxwell had shown that the electric and magnetic fields are manifestations of a single essence: the electromagnetic field.

1 Recall from chapter one that the *Système Internationale* designated unit of power is the watt and energy the joule, where a watt represents the conversion of a joule per second, and that a joule is 1 $kg\,m^2\,s^{-3}$ in MKS.

One consequence of his theory, among many, is the inevitability that electromagnetic (EM) waves would radiate from any disturbance. It quickly became apparent that light and radiated heat were merely EM waves which differ merely in their length, or frequency.[2] Radiation can be regarded as just another form of energy, transmissible through empty space, allowing the Sun to warm the Earth.

To make EM radiation, something with electrical charge has to wobble.

The American physicist Richard Feynman was once asked what shred of knowledge we should most strive to preserve should civilization fall. He replied "that everything is made of atoms". A star is no exception. While atoms do not have any net charge, their constituents do.

Electrons possess thousands of times less mass than the atomic nucleus. As a result, they may be expected to wobble a lot more — enough to account for almost all of the radiation we see.

When the atoms in a gas wobble about, we perceive it as 'hotness'. The more they wobble, the higher the temperature of the gas. As a gas gets hotter, its atoms eventually break apart, leaving a 'plasma', comprising 'ions' and electrons. Measuring the temperature of the Sun, or at least, its outer layer — the 'photosphere' — is not as hard as you might expect. Ever since a human first lit a forge, we have known that as something gets hot it begins first to radiate heat and then to glow, with a colour that changes with temperature. The process begins with a dull red and ends with white.

Colour is merely our perception of the wavelength of EM radiation, and thus of temperature.

It should not be surprising that temperature and power are strongly related. As something warms up, we might expect it to radiate more energy. A star can warm a planet across the void of space.

We've seen how EM radiation originates with the motion — to be more precise, the acceleration — of charged particles, like electrons. When that radiation collides with matter, there are two possibilities: either the ray is *reflected* or it is *absorbed*. In the latter case, the energy is converted back into heat, raising temperature. Most materials absorb some radiation and reflect the rest. Any surface which reflects everything is termed 'specular', after the Latin 'speculum' for mirror.

An object that absorbs *all* EM radiation falling on it is called a "black body". Rather more surprising is that the Sun constitutes a black body. It reflects nothing and is an almost perfect 'isotropic' emitter, radiating equal power in all directions. The relation between the total power radiated by the surface of a black body (per unit area) and its temperature is known as the Stefan-Boltzmann Law:

$$P_A = \sigma T^4 \qquad \sigma = 5.670 \times 10^{-8}\ \mathrm{W\,m^{-2}\,K^{-4}}$$

2 Wavelength (distance between crests or troughs) and frequency (number of complete waves per second) are inversely related — as the one increases, so the other is diminished.

After its spectrum, we now have a second way to estimate the temperature of the photosphere, since we can measure the radius of the Sun r_s quite precisely:[3]

$$L_s = 4\pi r_s^2 \sigma T^4$$

$$\therefore \quad T_s = \sqrt[4]{\frac{L_s}{4\pi r_s^2 \sigma}} = \sqrt[4]{\frac{3.85\times10^{26}}{4\pi\times4.84\times10^{17}\times5.67\times10^{-8}}} = \sqrt[4]{1.12\times10^{15}} = \sqrt[4]{1120}\times10^3 = 5{,}790\mathrm{K}$$

where L_s is the solar 'luminosity'. This is within a fraction of a per cent of the accepted value.

Of course, this estimate depends upon L_s, which we cannot measure directly. What we can measure directly is the irradiance of the Earth, above the atmosphere. This is sometimes presumptiously called the "solar constant", I_s. There is no reason to presume either the solar luminosity or the distance from the Sun to be exactly constant. For a start, the Earth's orbit is slightly eliptical, and is now believed to suffer long-term variation due to the gravitational field of the other planets, predominantly Jupiter. However, solar irradiance is now thought to have varied by only about 0.2% over the last four centuries.

To find the irradiance per unit area, all we need to do is divide the luminosity by the area of a sphere with the same radius as the Earth's orbit, R_e:[4]

$$I_s = \frac{L_s}{4\pi R_e^2} = \frac{3.846\times10^{26}}{4\pi\left(1.496\times10^{11}\right)^2} = \frac{3.846\times10^4}{4\pi\times2.238} = 1.37\mathrm{kWm}^{-2}$$

$L_s = 3.846\times10^{26}\,\mathrm{W}$

$R_e = 1.496\times10^{11}\,\mathrm{m}$

Next, we ask just how much solar power is available on the surface of the Earth. First, we multiply I_s by the area presented to the Sun by the Earth, then we divide by the area of Earth's surface:

$$I_{es} = \frac{I_s \times \pi r_e^2}{4\pi r_e^2} = \frac{I_s}{4} = 342\,\mathrm{Wm}^{-2}$$

Earth's mean radius[5] = $r_e = 6.371\times10^6\,\mathrm{m}$

Interestingly, this means we don't need to know the radius of the Earth, or any other planet for which we do the calculation. The area of a sphere is always four times that of the disc it presents.

We need to be careful with interpretation. It is still (theoretically) possible to recover close to the full 1.37 kW/m^2 by orienting the collector continuously towards the Sun. It will need to be tilted at an angle from the ground equal to the lattitude of the location, and rotate 180° every twelve hours. If the sky is clean, dry and free of clouds and other obstructions, we might get close. Sadly, no technology

3 We can easily measure the *angular* diameter of the Sun's disc. To estimate the linear diameter, we need to know the radius of Earth's orbit. For that, astronomers first made use of 'parallax' — the difference perceived in the position of the edge of the sun, against the starry background, at two locations of known separation. Today, we might gain greater precision by measuring the Earth's orbital velocity using the Döppler effect on the same starlight twice, six months apart.

4 If you wish to calculate something to three significant digits, you must employ parameters correct to four. Try the same calculation with the given parameters correct to just three digits and you will gain a different figure for I_s — 1.36.

5 Earth's equatorial radius is slightly greater at 6,378 km.

is anywhere near 100% efficient. Photovoltaic systems fall below 20%, while promising 40%, whereas thermal systems can exceed 70%, but are much harder to steer.

Without rotation and tilting, each dimension of the collector will be 'foreshortened' and the effective area presented to the Sun reduced, with correspondingly less power available.

All that energy from the Sun is either reflected or absorbed by the Earth. If the proportion absorbed went nowhere else, our planet would gradually become hotter and hotter. But, of course, hot things radiate. It's time to talk a little more about heat and temperature (hotness).

I can do no better than repeat the example presented to me by my physics teacher at school, Biff Bailey. Imagine a tank of hot water just right for the running of a bath — arguably, one of the greatest pleasures in life. We could measure its temperature with a thermometer and perhaps find that it's 70 °C. Now imagine a tank twice the size, but where the thermometer reads the same. We'd perceive no difference in our bath. The water in each possesses the same 'hotness' (temperature). However, it would allow running a *second* bath with the same hotness. Same temperature, twice the heat.

Since we revived the ancient Greek notion of atoms, we have understood hotness to simply represent their average velocity, as they race about inside their container. An obvious conclusion is that there may come a point, as the matter cools, where the atoms cease to move altogether, and that this must constitute a true zero on an *absolute* temperature scale. In physics, we use such a scale and name it after the great engineer and physicist William Thomson, who adopted the title Lord Kelvin. It makes use of the same unit (degree) as the centigrade (Celsius) scale, whose zero lies at 273 K.

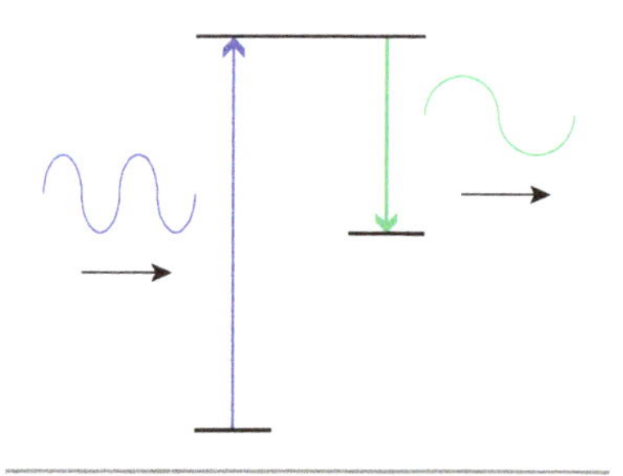

Figure 2.1 EM absorption/emission.

Heat is not found only in the linear motion of gaseous atoms or molecules. Molecular rotation and vibration can contribute to thermal capacity. Anyone who has ever placed their hands in front of a warm stove will be aware that heat can also be transmitted by EM radiation. All such radiation carries energy, but its absorption depends upon frequency. Atoms absorb mainly in the visible and ultraviolet, at a multitude of distinct values which depend upon their species. Molecules resonate in the infra-red, which thus earns the name "thermal radiation". They act like little aerials.

Atomic nuclei resonate only with radiation of *extremely* short wavelength (very high frequency), in the X and Γ (Gamma) part of the spectrum, and so hardly interact at all with solar radiation. It is electrons that absorb, and re-emit, visible and ultraviolet sunlight. Perhaps the most startling discovery ever was that they do so only at certain exact frequencies, called "spectral lines".[6] This is because an electron within an atom is permitted only certain distinct "energy levels" (Figure 2.1). The difference between these

6 The term "spectral line" follows the appearance of the spectrum within a spectrometer — an instrument for revealing a spectrum and measuring the frequency of EM radiation. A glass prism will suffice to reveal the spectrum of light, including dark lines where atoms in the solar photosphere have absorbed light. These can be used to identify the gas concerned.

is precisely proportional to the frequency of EM radiation absorbed or emitted. An electron is like a guitar string that has to be tuned with infinite precision for the guitar box to resonate at all. Once it has absorbed a visible 'photon' (quantum of EM energy), it can emit others of less energy and lower frequency, skewed by the Döppler shift due to motion. Because the velocity of atoms varies in both magnitude and direction, the spectral line is blurred a little and is observed with a finite width.

Earth's atmospheric gases are completely transparent to visible light. Solar energy is absorbed entirely by atoms within the surface of both land and sea. It is far from obvious how electronic energy is converted into heat, and so causes the surface to warm. Indeed, it wouldn't but for the ability of atoms to transfer energy from the electron 'orbit' to atomic vibration. Just like the electron, the atom is permitted only specific and exact levels of vibrational energy. A transfer can only take place when the gap between two electron energy levels precisely matches that between two of atomic vibration. In other words, the two systems are in *resonance*.

The vibration of one atom is easily transferred to neighbours, which constitutes one form of thermal conduction. Heat is also conducted when the vibration of an atom is converted to the vibration or rotation of a molecule, which might then lose it via thermal (EM) radiation.

Despite all the complex mechanisms whereby energy is carried from excited electrons to vibrating and spinning molecules, we can simply equate the power intercepted and absorbed by each square metre of the Earth's disc with that emitted by each square metre of its surface:

$$I_s \times \pi r_e^2 \times (1-\gamma) = 4\pi r_e^2 \times \sigma T^4$$

$$\therefore \quad \sigma T^4 = \frac{I_s(1-\gamma)}{4}$$

The Earth must be in "thermal equilibrium" with the Sun, if its temperature is stable.

To gain an estimate of the surface temperature, we need only determine γ, the proportion of incident energy reflected, which we call the planetary 'albedo'. But this is not so simple, because the surface of our planet is anything but homogeneous. Ice, ocean, forest, desert and (increasingly) concrete reflect very different proportions of incident EM radiation. Worse, the extent of the ice varies dramatically with season, and we have omitted the most variable component of all: cloud.

There is also the subtle question regarding exactly what constitutes the planetary surface. Because the atmosphere of our planet is so thin it will make little difference (to the value of r_e).

For the albedo, let's begin by using that of the Moon γ_m, which can be determined by straightforward observation, just like I_s, which we need to know in advance.[7] This seems reasonable now that we know that the Moon is composed of roughly the same raw material as the Earth.

---

7 We first observe the parallax of the Moon in order to calculate its distance R_m and thus its radius r_m. From this we can derive its irradiance. The radiance can be measured directly. Albedo is just the ratio of radiance to irradiance. →

$$T_r^4 = \frac{I_s(1-\gamma_m)}{4\sigma} = \frac{1.368\times10^3\times0.877}{4\times5.670\times10^{-8}} = 52.9\times10^8 \qquad \gamma_m=0.123$$

$$\therefore \quad T_r = \sqrt[4]{52.9}\times10^2 = 270\text{K}$$

Notice that we have no need to determine the size of the Earth to calculate its temperature.

So, if our world was as barren as the Moon then the *average* surface temperature would be –3 °C. It's important to stress that word 'average'. The tropics might remain habitable but little else.

The emergence of life would seem unlikely.

What distinguishes our planet immediately from the Moon is the presence of so much water. No one knows how it came to be here. It may have originated with comets, which regularly collide with the Earth, and accumulated over aeons. If the temperature were below freezing then we should expect a world whose surface is around 85% ice. Since dirty ice reflects about half its irradiance, and rock about an eighth, we might expect an albedo of about 0.45:

$$T_i = T_r\times\sqrt[4]{\frac{0.550}{0.877}} = 240\text{K} \qquad (-43\,^\circ\text{C}) \qquad \gamma_i=0.45$$

Clearly the figure employed for the albedo is important in deciding the equilibrium temperature of a planet. Given that something evidently warmed the Earth sufficiently for ice to melt and life to emerge, can we then account for seeing that warmth maintained? We can estimate a mean value for the Earth's albedo that accounts for the presence of ocean and forest, along with geographical, annual and seasonal variation. Accounting for the true nature and disposition of the Earth's surface:

$$T_w = T_r\times\sqrt[4]{\frac{0.840}{0.877}} = 267\text{K} \qquad (-6\,^\circ\text{C}) \qquad \gamma_w=0.16$$

So, a planet with an orbit and surface matching ours ought still to freeze.

Note that we have accounted for the effect of vegetation, which reflects more energy than rock, desert (much more), ocean (less, except at low angles of incidence), ice and snow (much more). We have also accounted for 99% of our atmosphere, comprising oxygen and nitrogen, which are almost transparent to the entire EM spectrum emitted by both Sun above and Earth below.

A little care is required here. Suppose the Moon behaved like a mirror. It would then be invisible unless the angle of incidence exactly equalled that of reflection. The opposite of so-called 'specular' reflection (after 'speculum', the Latin for mirror) is where incident radiation is scattered equally in all directions. A surface with that property is called 'Lambertian'.

The surface of the Moon is somewhat Lambertian, but varies significantly. It appears a little brighter when in opposition, which is partly explained by the loss of shadows cast by mountains. When measuring radiance, we must be sure still to average over both seasons and wavelength, accounting for the entire solar spectrum, including the infrared and ultraviolet.

What we have not yet accounted for are clouds, which reflect light efficiently. There are also aerosols, particulates and dust, but clouds are by far the most important. They almost double the energy reflected back into space, and are guaranteed on a warm world with so much water.

The true average albedo can nowadays be measured. Satellite measurements have brought some confidence here, along with those of lunar 'earthshine':

$$T_c = T_r \times \sqrt[4]{\frac{0.700}{0.877}} = 255\text{K} \qquad (-18\,^\circ\text{C}) \qquad \gamma_e = 0.30$$

Unsurprisingly, we find an even more inhospitable Earth, with an average surface temperature of -18 °C. This time, even the tropics might defeat us.

Clearly, we are missing something, but just how far out are we?

Measuring the thermal (infrared) emission of the Earth, again using satellite radiometry, might offer one way of estimating the true mean value of planetary surface temperature. There is also the obvious method where a great many thermometer readings are aggregated. These confirm a value of approximately 288 K (+15 °C), which is very good news for surface dwellers, such as ourselves.

So why is it so much warmer here than physics so far suggests?

Anyone who's ever watched a volcanic eruption has been confronted with the stupendous amounts of heat generated. Might this have warmed our toes, and allowed us to evolve?

The easiest way to dismiss this idea is simply to look at either planetary pole. Both the arctic and antarctic are frozen. We know from watching volcanoes, and from the transmission of sound through the ground, that a large proportion of our planet consists molten rock. If even a tiny proportion of that heat were to find its way through the solid crust over which we walk (or swim) then all that ice and snow would surely melt. It seems very unlikely that the poles are somehow uniquely protected.

The Earth's crust clearly forms an extremely effective insulator.

We must look elsewhere to explain our toasty home.

Application

2.1 The figures for albedo are given only to two significant digits, whereas everything else is given, or calculated, to three. This is because there is considerable uncertainty over albedo, which in any case varies both seasonally and (unfortunately) over time, as ice sheets melt.

Calculate the bounds for temperature under each of the above sets of assumptions.

2.2 The Earth's orbit is not a perfect circle but an ellipse. It is also prone to vary a little because of the gravitational influence of Jupiter. Assuming that its distance from the Sun can thus change by ±3%, calculate the effect you would expect on the Earth's surface temperature.

2.3 Calculate the expected surface temperature for both Venus and Mars under similar assumptions, given the following information:

	Venus	*Mars*
— albedo :	0.75	0.16
— orbital radius :	0.722	1.52 AU

where the Astronomical Unit (AU) denotes the mean radius of the Earth's orbit.

2.2 Pull up another blanket

All things convey heat to their surroundings as they warm up. The hotter they get, the more they do so. Otherwise, they would continue endlessly getting hotter. With the Earth isolated in a vacuum, it cannot evict heat by conduction. It can only do so via EM radiation — the same mechanism by which heat arrives from the Sun. The Earth will gradually emit more radiation until it gives up the same energy, per unit area and time, it receives. This is called "radiative equilibrium".

It might seem odd that a body as cool as the Earth can form any kind of balance with something as hot as the Sun. However, the Sun is a long way away, and the Earth receives only a tiny fraction of the energy it emits. From the solar perspective, the Earth is minute and lies in just one direction. To effect a balance, the Earth can radiate in *all* directions. Its rotation helps it lose heat and ensures an even temperature around its circumference. The land and sea that warm up during the day can radiate and cool down by night.

We have already seen one way how more energy can be carried by EM radiation : its frequency can rise, from the infrared, through the visible from red to violet, to the ultraviolet. Of course, a greater *intensity* will also carry more, which corresponds to a higher wave amplitude. The Stefan-Boltzmann Law expresses how the total power emitted per unit area is directly proportional to the fourth power of temperature. So, as a body warms up, it both changes colour and becomes brighter.

Thus far, we have assumed that all radiation emitted by our planet escapes into space. This would be true if the atmosphere comprised only oxygen and nitrogen, which are transparent in the infrared. But this accounts for only 99% of the gas above our heads. The remaining 1% contains a few things that change everything.

The following *absorb* infrared radiation:

— water vapour	H_2O	~1%	(variable)
— carbon dioxide	CO_2	360	parts per million by volume (ppmv)
— methane	CH_4	1.8	
— nitrous oxide	N_2O	0.3	
— ozone	O_3	~0.02	(variable).

As these gases absorb energy, they warm up. They too will begin to emit radiation until they reach radiative equilibrium with the ground below. This remains true, even though they will lose some heat by conduction to the other gases around them. This is exactly how a blanket works.

A blanket intercepts and absorbs some of the heat we radiate. If it is cooler than we are, it will radiate less onward than we would have done. It will also reflect back at us the radiation it did not absorb, while returning some it did. We therefore lose less heat overall, and our bodies need burn less fuel in order to maintain a healthy temperature.

An atmosphere differs from a blanket in one respect: while it affects the way energy flows out, it allows energy to flow freely in. This is known as the "greenhouse effect", because a garden greenhouse works in much the same way. An atmosphere acts like a glass roof.

Glass is transparent to sunlight, which therefore warms whatever lies beneath, which then emits thermal radiation. A proportion of this is absorbed by the glass, which itself then warms up and radiates both downward and upward. What lies inside the greenhouse sees the radiation from the glass *in addition to* that from the Sun. A balance outside the glass can be reached only when the temperature inside is significantly greater than it would be otherwise.

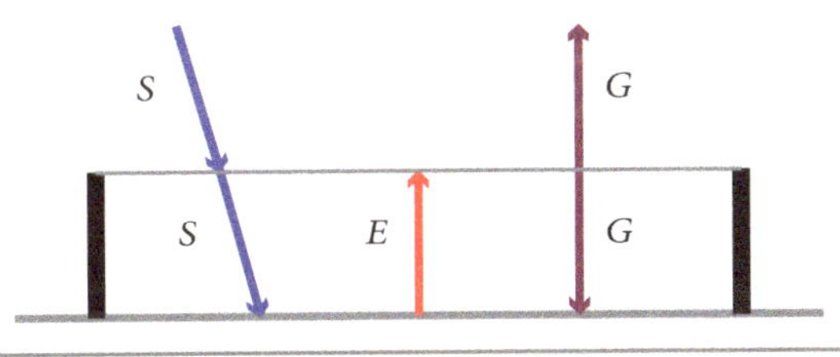

Figure 2.2 The simple greenhouse effect.

Figure 2.2 depicts an ideal system where the glass 1) is totally transparent to solar radiation, reflecting nothing, 2) absorbs all thermal radiation from below, and 3) is thin enough, or a sufficiently good conductor, that the temperature on top and bottom is the same. It will then radiate upward and downward in equal proportion.

Note the contrast between requirements 1 and 2. Glass just happens to have these properties. While it is famously transparent at wavelengths our eyes detect, it will pass only the 'near' infrared. (Ideal greenhouse glass would appear completely black at *all* thermal wavelengths.)

Given radiative equilibrium both inside and outside our greenhouse, we can see that:

$$E = S + G, \qquad G = S$$

$$\therefore \quad E = 2S$$

Without the greenhouse, E would just equal S.

To obtain an even warmer bed or planet, we can use more than one blanket, as shown in Figure 2.3. Provided it can absorb all the thermal radiation from below, each extra blanket has the same effect as the first, requiring anything beneath to radiate as much again. The number of effective blankets an atmosphere provides is called its "optical thickness" τ. If we begin with external irradiance E, the surface will end up experiencing $E(1+\tau)$. To account for reflection, we replace E with $I_s(1-\gamma_e)$.

The temperature will vary as the fourth root:

$$T_c^4 = \frac{I_s(1-\gamma_e)}{4\sigma}, \qquad T_g^4 = \frac{I_s(1-\gamma_e)(1+\tau)}{4\sigma} \qquad \therefore \qquad T_g = T_c\sqrt[4]{(1+\tau)}$$

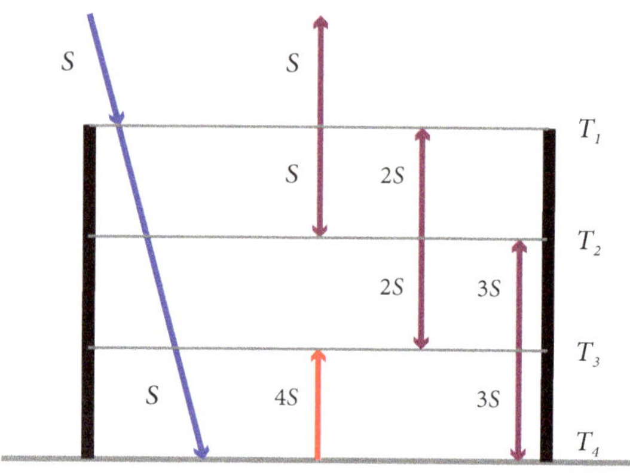

Figure 2.3 The compound greenhouse effect.

An optical thickness of one will thus double surface irradiance and increase temperature (on the Kelvin scale) by the fourth root of two, or about 19% — a rise of more than 48°C. From satellite measurements, we can infer an average value for τ of 1.87. If the greenhouse effect were all that mattered, we would be forced to expect an average surface temperature of about 59 °C, even after accounting for the loss of nearly a third of the Sun's incident energy by reflection back into space.

$$T_g = T_c\sqrt[4]{(1+\tau)} = 255 \times \sqrt[4]{2.87} = 332\text{K } (59\,^\circ\text{C})$$

Again, though some creatures might survive exposure to such an environment, life would be mainly restricted to mountain tops and the deep ocean. Some planetary explorers, who work thus far by robotic proxy, propose that life might arise floating in the sky. Temperatures decline rapidly with altitude in our current model, as you rise above each successive 'blanket'.

Perambulating apes would have a tough time indeed.

Whenever we obtain a result which is out of kilter with what we observe, the first thing (after checking our work) is to re-examine every assumption on which our calculation was predicated. Foremost is that the atmosphere is completely transparent to all solar irradiance. Although the vast majority of solar emission falls within the visible part of the EM spectrum, some remains at longer wavelength (lower frequency), and can thus be absorbed by 'greenhouse' gases. In fact, about 20% of the total irradiance is absorbed by the atmosphere and clouds.

Neither is the atmosphere completely opaque to the Earth's emission. About 4% escapes directly out into space. Our planet has the equivalent of two *imperfect* blankets.

Figure 2.4 depicts the flow of radiative energy within our model thus far. Though it may seem absurd, our atmosphere is warmed more than twice as much by the Earth than the Earth is by the Sun.

The effect on the surface irradiance of atmospheric absorption of solar radiation is the same as an increase in average planetary albedo from 0.3 to 0.5:

$$T_g^4 = \frac{I_s(1-\gamma_e)(1+\tau)}{4\sigma}, \quad T_g'^4 = \frac{I_s(1-\gamma_e')(1+\tau)}{4\sigma}$$

$$\therefore \quad T_g' = T_g \sqrt[4]{\frac{(1-\gamma_e')}{(1-\gamma_e)}} = 332 \times \sqrt[4]{\frac{0.5}{0.7}} = 305\text{K} \quad (32\,^\circ\text{C})$$

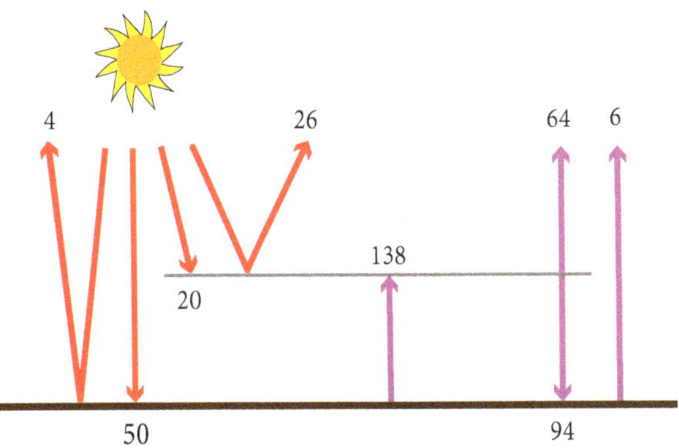

Figure 2.4 Earth's radiative energy budget.

While this yields a cooler surface, where apes might perambulate in temperate latitudes, sweating profusely, it does not account for the 15 °C average that we actually see.

Something else is clearly going on.

2.3 Long may it rain

Things are a little more complicated still, thankfully.

We've just seen that only half of the Sun's energy reaches the ground. Almost half of that is given up in the evaporation of water. All that water must come down again, but *not the heat*. As it rises, the vapour cools and condenses back into liquid droplets. Just as evaporation takes up energy, condensation releases it. The net effect of evaporation and rain each year is the transport of a gigantic amount of heat from the surface into the atmosphere, at a wide range of altitudes. Some of that energy is then radiated out into space, restoring balance, but, at all but the lowest layers, the atmosphere is rendered warmer than it would otherwise be. The surface is made much cooler.

Occasionally, and rather badly, I fly gliders — one reason why sheep are nervous in Oxfordshire. To fly a glider for more than a few minutes, you have to find a 'thermal' — a column of rising air, warmed by the ground beneath. What goes up comes down. Find the warm rising air, you go up; find the cool falling air, you go down. Experienced glider pilots can fly hundreds of miles in a single hop using just thermals, although there are other solar powered phenomena they can employ.

A thermal is an example of oddly named 'sensible' heat, collected from the ground by conduction and transported up into the sky by convection. Air circulates in both the horizontal plane and the vertical. Again, the atmosphere is warmed, and the surface cooled.

The influence of clouds is far from simple. On their own, they reflect about a fifth of the incident solar energy. They also absorb and emit thermal radiation, and so add an extra blanket when they are around. (A cloudy night is usually warmer than it would otherwise be.) These two effects counteract, making clouds difficult to account for in any model. One cools the surface, the other warms it. The net result is believed to be a slight cooling. Vapour trails, left by airliners, act the same way.

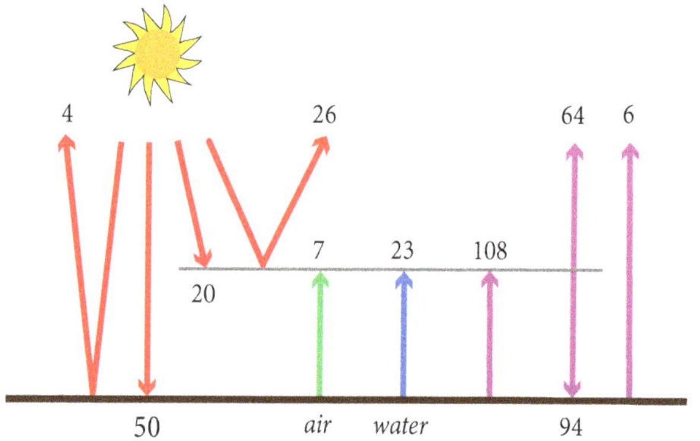

Figure 2.5 Earth's energy budget, accounting for weather.

Figure 2.5 depicts the solar energy budget of our planet and conveys something of the precarious balance on which life depends, particularly on land. As well as absorbing a fifth of the incident solar energy, plus most of the thermal radiation from the ground below, the atmosphere gets the heat brought by convection and latent in evaporated water.

Note that the heat transferred to the atmosphere from the ground by air and water does not change the equilibrium between Earth, Sun and space. It merely *replaces* that which would otherwise be radiated by a warmer surface.

Life on our planet, on land at least, is saved by the presence of *weather.*

The Stefan-Boltzmann relation can be used once more to confirm that the energy transfers due to weather shown in Figure 2.5 are enough to explain our comfortable home. We first refine our estimate of the total surface irradiance (energy arriving) *without* weather:

$$E_g' = \frac{I_s}{4}(1-\gamma_e')(1+\tau) = I_{es} \times 0.5 \times 2.87 = I_{es} \times 1.435$$

This figure (rounded) equates with the total surface radiance (energy leaving).

If we subtract the thirty units ($0.30 \times I_{es}$) given up through evaporation and convection, we can compute the temperature the surface must have in order to radiate the remaining energy upwards:

$$E_w = I_{es} \times 1.135$$

$$\therefore \quad T_w = \sqrt[4]{\frac{I_{es} \times 1.135}{\sigma}} = \sqrt[4]{\frac{342.5 \times 1.135}{5.670 \times 10^{-8}}} = 288\text{K} \qquad (15\,^{\circ}\text{C})$$

By transporting about a fifth of surface energy up into the atmosphere, through evaporation and convection, our weather does indeed account for the average planetary temperature we see.

We should be very glad of rain. As we have seen, evaporation of water is largely responsible for cooling the surface of Earth. There is another reason, however.

Now that our robotic progeny have visited it, we know there is no Mekon on Venus, no Treens to attempt our conquest, no Therons to make good our waning food supply and no belt of fire for Dan Dare to brave. Shame really. Instead, there is an atmosphere a hundred fold thicker than our own, through which very little sunlight can penetrate. What is more, three quarters of the solar energy incident on Venus is reflected, compared with less than a third by Earth. Though one might expect the surface to be therefore cool, a temperature of 525 °C has been recorded — high enough to melt lead.

One sufficient reason for space exploration is to learn more about our own world by studying others. No amount of theory can ever substitute for observation. Venus and Earth are of similar size, share a similar position in the solar system and are likely to share the same chemical composition. We might well ask why Earth has not already suffered a similar fate, and whether it could still do so.

We can begin by assuming the same starting conditions: each planet possessing an ocean, but only a thin atmosphere. However, Venus is a little closer to the Sun and thus receives more solar energy. Evaporation will have been more rapid. Here we have the nub of the issue: water vapour is a *very* powerful greenhouse gas. The consequent greenhouse effect would have further warmed the surface, encouraging more evaporation, and so on until either all the ocean had become vapour or the atmosphere had become 'saturated' (*i.e.* could hold no more).

In fact, the rapidly warming air of Venus would *never* have become saturated. A 'runaway' greenhouse effect thus sent its temperature soaring, limited only by the quantity of water available. On the cooler Earth, saturation occurred early on. No runaway effect was ever possible here. Evaporation is free to continue cooling the surface indefinitely. The result of any excessive evaporation is just rain.

It can never rain on Venus.

Life is clearly *very* delicate. Without an atmosphere at all, our world would be too cold. With one, but without weather, it would be too hot. As it is, we have it just about right. One might wax Darwinian and suggest that the only reason it is about right is that we have evolved precisely for that temperature. Were it warmer or cooler, we would have adapted and it would still be about right from the new perspective.

Unfortunately, now that we are able to go and look at some other worlds, we find no obvious signs of life in any different environment. True, we have yet to look everywhere. Europa and Titan are still under investigation, but results thus far do not augur well. Most of those searching would be over the Moon (pun intended) to find even microbes. As yet, we have no reason to believe that life, still less *macrobiotic* life, could adapt to any temperature much above or below that we are used to.

We should not confuse the effects of diurnal or annual variation, which we successfully manage, with those of a significant change to the *average* ambient temperature. More than a 10 °C variation from the norm, and our very survival is threatened.

It may take a lot less to cost us our civilization.

Application

2.4 Assuming no clouds and no weather, what greenhouse would warm Mars?

2.4 The burning question

Most stars appear to shine with remarkable constancy. Precise measurements of solar luminosity only extend back about thirty years but indicate a variation of less than one tenth of a per cent. Extrapolations, correlating solar activity with output, suggest that up to half a per cent variation may have occurred over the last three centuries. A cool period in the 17th Century may have been a consequence.

On the other hand, we have long known that the orbit of the Earth wobbles. We can reliably calculate that, for the last million years or so, its eccentricity increased a little every hundred millennia, bringing the planet a fraction closer to the Sun twice a year.

Growth rings in tree and coral can reveal the climate of the recent past, as can pollen found in glacier, ice-cap and sediment below lake and ocean. The sedimentary pollen record extends back around fifty thousand years. Beyond that, we can resort to fossils of microscopic sea creatures. Ice core samples can reveal the carbon dioxide (CO_2) concentration over at least four hundred thousand years.

The climate record tells us that average surface temperature does indeed increase when the Earth ventures nearer the Sun. In between warmer periods are "ice ages", when glaciers extend down to European latitudes. The last one began about 70,000, and ended just over 10,000, years ago.

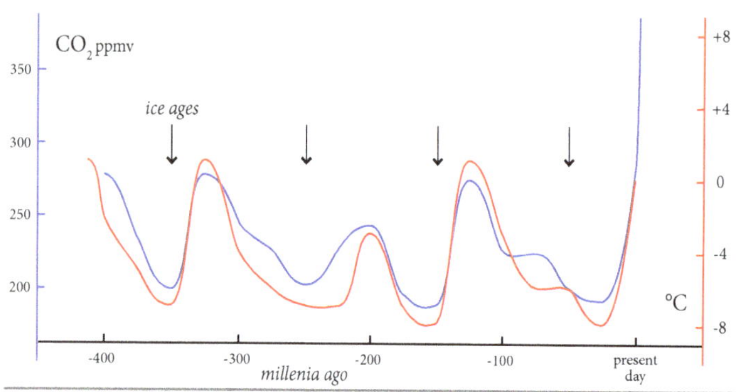

Figure 2.6 Temperature versus CO_2 concentration over millennia.

A very small change in solar input clearly brings about a *major* climatic response. Radiative equilibrium alone cannot explain such a dramatic reaction. Something else must play a part.

Ice reflects sunlight very efficiently. As glaciers grow, more energy is lost in reflection, causing further growth, and so on. If we overlay the records of surface temperature and atmospheric carbon dioxide concentration we see a strong correlation (Figure 2.6). It is reasonable to believe that a change in the Earth's orbit caused a change in temperature, and that this in turn caused the variation in CO_2 and thus in the degree of greenhouse effect. We call such amplification "positive feedback".

Not all feedback is positive. Some responses tend to *quell* a change. The importance of feedback is borne out by attempts to reproduce the temperature variation seen in the ancient record using computer models. These make plain the interacting feedback loops of a "complex system".

Glaciation and a reduced greenhouse effect alone are sufficient to explain ice ages.

Temperature leads; greenhouse gas concentration follows.

But can it work the other way around?

Atmospheric carbon dioxide concentration has leaped upward by 50% since the start of the industrial revolution, and is rising rapidly. In and out of an ice age, it varied between 180 and 280ppmv,[8] but is now up at around 380, and is increasing by at least 1.5 each year. This corresponds to an annual addition of approximately 3.3 gigatonnes (GT) to the atmospheric carbon reservoir.

Clearly, something dramatic, perhaps seminal, is occurring. There can now be no reasonable doubt that it is a result of our burning fossil fuel. The change just correlates too well with increasing industrial activity. A strongly correlated rise can be seen in all three primary greenhouse gases, aside from water vapour: carbon dioxide, methane and nitrous oxide. One can even make out two kinks, matching post-war acceleration of production (Figure 2.7).

Temperature can be clearly seen *following* greenhouse gas concentration.

Modelling the climate is now popular internationally and very well funded. Nations compete vigorously to host the most powerful computer applied to this end. No model has succeeded in reproducing the rise in temperature thus far seen without including anthropogenic emission — the result of our burning gigantic amounts of fossil fuel. Two thirds of the evident enhanced greenhouse effect may be attributed to carbon dioxide emission and about a quarter to methane, which is the dominant constituent of natural gas.

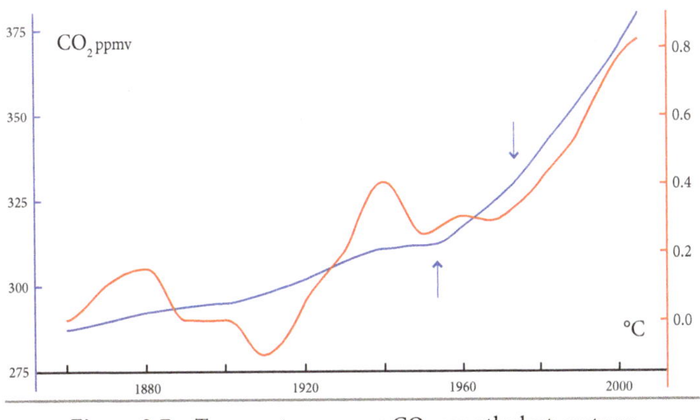

Figure 2.7 Temperature versus CO_2 over the last century.

The average surface temperature on our planet rose by 0.6 °C during the 20th Century. Computer models indicate an additional increase over the 21st Century of somewhere between 1.4 and 5.8 °C. The United Nations International Panel on Climate Change (IPCC) place their current best bet on 2.5 °C. This comprises 1.2 °C as a directly enhanced greenhouse effect plus 1.3 °C as a

8 Parts per million by volume.

consequence of feedback. Some feedback effects are already in evidence, such as the rapid melting of glaciers and the arctic ice-cap.

These numbers do not sound large or frightening in themselves, but it must be remembered that they are *averages* over the seasons and all latitudes. A little thought regarding the energy involved is perhaps more sobering. To put it another way, we are trapping energy equal to that liberated by detonating a standard 10mt thermonuclear warhead every twelve seconds.

It is now only a matter of two or three decades before the pre-industrial (but post ice age) atmospheric carbon dioxide concentration is doubled. This reduces the rate at which the surface can radiate heat into space by about 2%. Again, it doesn't sound much, but we are dealing with quantities of energy that make our paltry consumption pale into insignificance. That 2% amounts to around 3,600 TW — *three hundred times* the 12 TW we humans consume worldwide.

One might hope for mitigation. After all, some of the feedback known is negative. For example, on land, plants are fertilized by an increase in carbon dioxide, and so digest more of it. Some measure of regulation arises this way. Unfortunately, there has been extensive tropical deforestation to add to that which had already taken place in the northern hemisphere (where the greater part of the land mass is found). Although some re-establishment of northern forest is now being fostered, something like an additional annual gigatonne of carbon remains in the atmospheric reservoir.

Our familiarity with carbonated drinks might lead us to think that the ocean would absorb some of the excess. Our species actually liberates around 7.5 Gt of carbon each year — 6 Gt from burning fossil fuel and 1.5 Gt from changes in land use and deforestation. Of this, around 45% (the 3.3 Gt mentioned earlier) remains in the air. The rest does indeed end up in the sea, which cannot do better because only the surface layer (the first few tens of metres) can be effective in the short term. Mixing of these waters with lower layers takes centuries, or even millennia for the great ocean deep.

Fertilization of oceanic life may constitute another possible negative feedback. There is some evidence from the ice-core paleoclimatic record that the plankton population did increase at the end of the last ice age, as both temperature and carbon dioxide levels rose. This would have pumped more carbon out of the atmosphere and into the ocean. However, there is no direct evidence yet reported of that effect in modern times.

While no other significant negative feedback is immediately apparent, no fewer than three further positive feedback loops are predicted. Higher temperatures affect the respiration of land life, especially soil microbes, increasing their carbon emission. The stress of climate change is expected to slow forest growth, reducing atmospheric carbon 'capture' and 'sequestration'. Finally, as warming proceeds, there comes a point where methane is released, first from wetland areas, and second, and of greater long-term concern, from huge reservoirs of methyl hydrate in deep ocean sediment.

Some of the worst predictions are coming true decades before even alarmists expected. The perennial arctic sea ice is melting. An open sea reflects far less solar radiation than an ice sheet — positive feedback of the first order. Siberian permafrost is melting, liberating vast amounts of methane.

It seems pretty clear that we as yet have no reason to expect salvation from nature, and every reason to expect feedback to exacerbate climate change. As we learn more, our uncertainty actually grows.

Perhaps the most disturbing news is that the change in atmospheric carbon dioxide concentration is accelerating. Over the final half of the 20th Century, the average annual rise was just 1.3 ppmv; over its last two decades it was 1.6; but over the first fifteen years of the new century it lies in excess of 2.0.

Even with our new-found computer technology, building a model that accounts for all known feedback is extremely difficult. One outcome is an upper and lower bound for the Earth's surface temperature at the end of the 21st Century. While the lower estimate concurs with earlier ones, the higher one predicts a rise of up to 11 °C. No one knows all the consequences of a rise that large.

One we can be sure of is a lot more wind and rain.

Application

2.5 Why do both Venus and Mars have a far higher concentration of atmospheric carbon dioxide than Earth?

2.6 Suppose the rise in concentration of greenhouse gases causes the optical thickness of the atmosphere to increase to 2.0, but more violent weather causes a corresponding (10%) increase in heat transported away from the ground. What rise in surface temperature would you expect?

2.5 You versus the universe

There was a time when we knew security only when at peace with a god, to whom everything that happened was directly attributed. When someone died, your house blew down or a conquering army arrived to destroy your entire community and culture, you knew you had offended a deity.

Truth itself was defined by the word of God, as related by someone unknown and long dead.

Perhaps our greatest struggle today is between those who interpret their word in a most literal sense and those who choose greater abstraction: the 'fundamentalist' versus the 'liberal'. Maybe there is a greater *sense* of security in the former perspective, and certainly less room for argument. But plague and tempest still wreak havoc, while the slightest departure brings inquisition.

Great claims are made for science: that it replaces fearful superstition with confident objectivity; that it redefines 'knowledge' in a way open to all; and that *everything* is open to explanation, through careful observation and analysis. We live in an age where these things are widely accepted, by the liberal at least, if not the fundamentalist. For some, science has completely replaced religion.

Only a fool would challenge the benefits that science has brought via the technology it makes possible. We have a fridge to keep food fresh, telecommunications to keep us in touch and aviation to travel the world — all affordable to everyone partaking in a developed industrial economy. Our security depends on our having better technology than our enemies. Arguably of far greater value though is our ability to explain everything we encounter from day to day. It is quite impossible for us to comprehend the fear experienced by our ancestors in the face of even quite ordinary events.

Even if we are at a loss ourselves, we feel quite certain that *someone* somewhere understands.

Just as it is almost impossible for us to imagine life without technology, we shall never know the power the clergy once held over us, as the keepers, guardians and final arbiters of sacred truth. The world's first scientific society — *The Royal Society*, in London — adopted the motto "Nullius in verba", which is a reference to the work of the Roman poet Horace:

> "...nullius addictus iurare in verba magistri..."

> "...binding on no one to swear by the words of any master..."

This is often mistranslated as "take nobody's word for it", which carries the same sentiment: to deny the authority of any individual, when it comes to truth. The only thing sacred in science is that *nature* shall be the final arbiter. Observation is thus sacrosanct.

We can all look for ourselves.

Throughout history, each civilisation is ultimately built on two things: slavery of some kind, to elevate production sufficiently to allow at least a minority to indulge in philosophy and the arts, and some set of unifying beliefs and values which, along with language and culture, dictate identity. For the last two centuries, throughout the developed world, science has supplied each of these requirements. Technology has already supplied machines which have relieved about a third of the need for labour, while increasing production many fold. It perhaps has the potential to replace labour altogether, if not enterprise. And belief and faith in science has to a large extent displaced religion.

Unfortunately, it shares the same potential for corruption and incompetence as any church.

First, there is a limit to its intrinsic competence, as well as that of its practitioners — the new priesthood. For some reason, we teach science in our schools without any acknowledgement of what we *don't* know, leaving every child to believe the lack of an explanation for anything is their *own* failure, rather than that of their society. This is particularly sad since what renders science most fascinating is precisely the sense of wonder in those things which are not yet fully understood. The sin extends to giving improperly substantiated explanations. Sadly, "nullius in verba" should extend to everything one reads in textbooks, since they are all too often wrong. If you enjoy dusty old second-hand-book shops, you should seek out old texts on, for example, astronomy. As you progress in time, you will read first that the Sun is fuelled by coal (it would collapse under its own weight and burn out after only a few centuries), then that it's powered by nuclear fission (it would generate huge numbers of neutrons we simply don't see). Even the current explanation — nuclear fusion — suffers at least one inconsistency. The Sun would generate a huge neutrino flux, to which we do not (yet) bear witness.

And here is a second problem for the authority of science: observation is rarely as simple as looking out the window. It is often extremely difficult and vulnerable to error. Science has an answer to this problem: to be accepted, an observation has to be *repeatable*. Someone else must verify it, preferably using a wholly distinct method, under different circumstances. Unfortunately, this response is not entirely satisfactory. Any subsequent observer will be biased. Depending on their private persuasion, they will either favour repetition or denial. Either way, a bias arises.

It will also be difficult to repeat an observation if it requires a billion-dollar instrument, or a network of a million accurate and calibrated thermometers evenly spaced around the Earth.

The result of all this is the unintended promotion of unelected individuals to the rank of high priest. Their hat says 'scientist', like the old pointy one said 'bishop', and we're expected to believe what they say because of that. Don't. To be true to science, they must *show* the truth, not dictate it.

Scientists are just people conducting an investigation. They are human, and are thus open not just to incompetence but also to all the usual human failings. Contrary to some newspaper reports, there is little money to fight over; at least, little that might ever end up in their pocket. But there is *reputation* — the esteem in which their peers hold them — which is worth a lot more to some than cash.

Success in a highly competitive world is rarely down to passion or ability. Enterprise, leadership and social, communication and organisational skills are as important as they are in business. You may be brilliant but if you can't bring in the research grant and write the paper, you won't get the job.

We should not be surprised if the big cheese sometimes fails to measure up.

An example can be found in the story of fat in our diet. A very senior figure in the 1950s US medical community alleged that excessive consumption of fat was the leading cause of heart disease, which had risen dramatically after the war. Because of his rank, a huge campaign to reduce fat in our diet was mounted, on both sides of the Atlantic. It successfully reduced fat intake by about a third, but the incidence of heart disease continued to rise, unabated. We now know that there was no significant evidence that fat was ever to blame. A high priest, not science, had spoken.

Thankfully, there is an obvious counter-example, which brings great credit to the world of science. In 1905, a lowly patent clerk, with neither rank nor sponsor, submitted five papers to a leading journal. Despite his lack of either academic connection or established reputation, they were accepted for publication, even though each was somewhat radical, even contentious, at the time. All were to prove of a standard sufficient to gain nomination for a Nobel prize. His name was Albert Einstein.

One wonders, sadly, how many Einsteins have been excluded (and plagiarised) since.

A negative result of Einstein's success and consequent celebrity has been the elevation of theory over observation in physics. Observation must test theory, as well as inspire it, but this should play a secondary rôle since, once again, a bias is introduced. Most valuable by far are those experiments whose outcome remains completely unpredictable, and where expectation is entirely balanced.

We are not yet done with the limitations of science, in replacing religion in our lives. There are two more concerns. First is the recent discovery of the behaviour of some 'complex' systems, termed 'chaotic'.

The technology underlying the industrial revolution depended upon an understanding of both thermal and mechanical systems, afforded by progress in physics during the 18th and 19th Centuries. The success of both the science and the consequent technology gave rise to a common perception of the world as one big machine. The only great surprise was in the discovery that the conversion of thermal into mechanical energy and vice versa could not go on forever. Energy was 'degraded', posing new questions concerning the origin and fate of the universe. But at least the grand machine was *predictable*.

All you needed was enough information about the beginning of the universe, and everything in it, and you could calculate all that happens right up to its demise. Any uncertainty at the start would just give rise to a similar degree of uncertainty later.

The truth was there to see in the 1960s, though it took another two decades before it was understood. Some systems sometimes behave in a way that is *totally* unpredictable. Infinitessimal uncertainty at the start leads to infinite uncertainty later. Our weather is prone to this. A common illustration is how the beat of a butterfly's wing on one side of the world might lead to a tempest on the other. It simply isn't possible sometimes to collect enough information to make a forecast.

No science will ever afford certainty in answer to every question, no matter how simple.

Lastly, while science can often tell us *how* something happens, it can never tell us *why*. That lies outside its domain, leaving valuable territory open to personal belief and religion.

Physics seeks to reduce the number of entities necessary to explain all phenomena observed. Philosophers argue as to what that final number might be, to remain logically consistent. Perhaps everything we see is the consequence of disturbance to a single universal field. Perhaps at least two such entities are necessary in order to suffer interaction, thereby giving rise to all that we see.

One thing remains certain. The absence of any such entity, and thus all experience, can break no rule or law. And yet here we are, complete with stars and galaxies, spinning and evolving over æons. We inhabit a universe which did not need to exist.

That, in itself, does not imply creation or the existence of a creator. The universe may exist in perpetuity or be cyclic in nature. Science may yet resolve such questions.

But it will never explain *why* we are here, and here is here.

Savour the mystery!

Chapter 3

Space

3.1 Onward and outward

The Earth, our cradle

For most of us, the term 'world' refers to our planet. Any extension of that reference would normally be condemned as fantasy, and might even be met with laughter, despite footprints on the Moon. After all, the Apollo program cost hundreds of billions of dollars and is now consigned to history. However, we should remember that aerial flight was once dismissed as likely impossible, extremely expensive and dangerous at best. Yet now, even the poorer among us regularly fly abroad just for a holiday.

Technology can turn economics upside down.

Our descendents may well shudder at the thought of living on a planetary surface. We have just seen one good reason why. In all honesty, there is no chance whatever of decisive action being taken to protect the climate against an increasing greenhouse effect. Politics is powerless against either rising population or rising consumption. It is also powerless against the vested interests of those who make vast profits from both deforestation, for timber and agriculture, and the burning of fossil fuels.

There will be many more hurricanes, typhoons and storms, of increasing severity.

Weather is only one from a range of traditional threats we face from our home. There is little technology can do to protect us from volcanic eruption and earthquake. We can enjoy much better monitoring and prediction, of all three, but a bitter toll in death, injury and destruction remains. Few of us bear space in mind, but we are sitting ducks for asteroids and comets with whom we share our solar system. Around twice each century, on average, the Earth suffers the impact of an object large enough to destroy a city. We are slowly gaining the capacity to track asteroids, but cometary orbits are much less predictable, as they evolve as they approach the Sun. Through history, the threat was small because humanity was spread far and thin, but now we live everywhere on land. An impact at sea would cause a tsunami, and so still threaten many cities, which we prefer to build on the coast.

As our population and consumption rise, another threat emerges: depleting resources. Much is made of our depleting energy reserves, by which most reports mean fossil fuel. Even if we do run out of oil, coal and gas, there is in fact a plentiful and *sustainable* supply. We globally consume around thirteen teraWatts, yet the Earth absorbs around one hundred and twenty thousand from the Sun. There is also enough energy generated beneath our feet to melt the entire planet, bar the shallow skin on which we walk and construct our homes. The problem is getting it to where, and when, it's needed.

There's a lot of it about but it's hard to store and carry, which explains the appeal of oil.

Aside from creating greenhouse gases, burning fossil fuels also destroys a vital 'feedstock'. For example, oil is used to make plastic — a rather important commodity — which leads us to the real problem: we are running low on other resources vital to satisfying global demand.

Arguably the most destructive demand is for food. Agricultural production is the dominant cause of deforestation. In newly prosperous parts of the world, entire populations are converting to a carnivorous diet, with a strong preference for red meat. Producing beef requires nearly *thirty times* more land than pork or chicken, and more than a *hundred and fifty times* as much as legumes or grain.

It also consumes ten times as much fresh water and liberates ten times the greenhouse gas.

The only hope is in recent trends which follow the arrival of prosperity. Education, especially among women, has led to a fall in birth rate. People still have children, just fewer and later, which in turn advances their chances for a better life, dramatically. Also, people in rich countries seem to opt out of ever-increasing consumption as they grow older, preferring to retire earlier, or at least work less.

Sadly, even when we take account of these changes, our planet faces a poisonous threat due to increasing population and demand, which is expected to peak at perhaps ten times current levels.

We are fast outgrowing our cradle.

The challenge of space

The politicians who fund manned space flight have, no doubt, traditionally justified the expense in terms of national prestige or military prowess. An enemy could easily tell that a rather large bomb could replace the little spacecraft in which astronauts sit. Few gave thought to other implications.

Each successive space ambition requires the solution of at least one key problem. To put a person in orbit around the Earth, or to escape the clutch of its gravity altogether, simply requires the ability to hoist their mass high enough and fast enough. As we will see later, the "big dumb booster" rocket is hardly a satisfactory solution for anything more than a stunt. It is incredibly dangerous, expensive and inefficient. Only the bravest and fittest individuals are accepted as astronauts.

But a proper solution would also solve a serious problem down below: how to traverse the longest distances around our planet both faster and more cheaply. A ballistic trajectory could take you from London to Sydney in a small fraction of the time taken by a jet airliner.

The same general principle holds for each of the technological problems facing spaceflight, and forms a far better justification for the huge expense in research and development.

To visit the Moon, we had to learn to *recycle air*. In 1970, after an explosion in a power module, the lives of the Apollo 13 astronauts were threatened most directly by the failure of this system. Only the greatest ingenuity saved them. Down here, a fast-growing problem is air pollution, which is believed to kill around forty thousand people each year in Britain alone. While you might feel safe in your private car, you are actually at greatest risk there. Of course, a more sensible solution would be to replace the internal combustion engine (particularly the diesel one) with an electric motor. In the meantime, while we regain our sanity, air recycling would be valuable in both car and home.

It is also necessary in submarines, and potentially valuable in airliners.

To build a space station, you must efficiently *recycle water*. Water is heavy stuff, so carrying an adequate supply is going to be very costly, and will limit the number of astronauts living there. The International Space Station (ISS) recycles water with about 93% efficiency. Down on Earth, water may seem in plentiful supply, but that is not the case everywhere. In many parts of the world, fresh water has become expensive and may even come to threaten war between neighbouring countries.

Recycling has to be a better option than laying thousands of miles of pipes and drains, if it can be done efficiently, and using only a modest and sustainable energy supply.

To go to Mars, there are two huge problems to solve. By far the greatest is how to efficiently *recycle food*. Nature accomplishes this on Earth, but only by deploying an extensive ecology and an enormous area, neither of which is available on a spacecraft. Yet to succeed here would guarantee an adequate supply of food to the entire global population, and without deforestation. Any candidate process would have to accomplish the feat using only a limited supply of solar energy, which is also available everywhere on Earth. As well as feeding the hungry, it would relieve the pressure on our ecology.

Of course, it would be far simpler to agree only to bring a child into the world where and when the resources are available to feed them properly. But there is no chance of any such agreement.

Finally, as our landfill sites overflow with things we no longer need or want, we must learn the art of *'closed-cycle' manufacturing*, where products return to the factory at the end of their life. The design of each one simply incorporates consideration of its end, as well as its beginning.

There will be no trash pits in space, or (hopefully) on other worlds we colonise. Yet there is an almost infinite supply of every resource, including directly accessible energy.

Almost all technological advance has arisen through war or the threat of such. For the first time in history, we have the opportunity to challenge ourselves in a more productive manner. In exploring other worlds and building habitats in space, we can also understand our place in the universe.

We can redefine our world.

3.2 Leaving the cradle

Escape

Certainly the greatest technological barrier to space currently is getting there, even though it isn't very far away. It lies just a hundred kilometres above our head. We begin by asking why gaining altitude requires so much work (*i.e.* energy).

Of course, if we could just get in a lift and ascend above our atmosphere, we would find ourselves in space. Unfortunately, no conventional material could withstand the weight of a tower high enough, but it's worth noting the additional energy required, per kilogramme: [1]

$$E_h = mgh \simeq 1\times10\times10^5 \text{ joules} \qquad g \approx 10\,\text{m/s},\ h \approx 10^5\,\text{m}$$
$$\simeq 1 \text{ MJ}$$

Were we to throw that 'payload' vertically upward, we would thus need to impart a velocity:

$$\tfrac{1}{2}mv^2 = mgh$$
$$\therefore \quad v = \sqrt{2gh} = \sqrt{10^6} = 10^3\,\text{m/s}$$
$$= 1\,\text{km/s}$$

Of course, it won't then stay up there; it would return to land on our heads with the same velocity with which it began. At 'apogee' (the furthest point from the planet), we might once again somehow accelerate the craft, parallel to the ground, so that it can sustain a circular orbit: [2]

$$F_c = \frac{mv^2}{r} = F_g = \frac{GM_E m}{r^2} \qquad mg = \frac{GM_E m}{R_E^2}$$

R_E = radius of Earth >> h

$$\therefore \quad v_{leo} = \sqrt{\frac{GM_E}{r}} = \sqrt{\frac{GM_E}{R_E^2}\times\frac{R_E^2}{R_E+h}} \simeq \sqrt{gR_E} \approx 10\,\text{km/s}$$

It has become apparent that the velocity needed to obtain even a low orbit around the Earth is at least ten times, and the energy *fifty* times, greater than that to merely gain altitude. However, if we launch from somewhere near the equator, the planet's rotation makes a modest contribution: [3]

$$v_{eq} = \frac{2\pi R_E}{1 \text{ sidereal day}} = \frac{2\pi\times6.378\times10^6}{8.6164\times10^4} = 465\,\text{m/s} \approx 0.5\,\text{km/s} \qquad R_E = 6.378\times10^6\,\text{m}$$

It may not seem much but every little helps (and saves money).

---

1 See Chapter 1, Section 1. 3.6 MJ = 1 kWh, which costs an electricity consumer around 10 ¢.

2 See Chapter 1, Section 2.

3 A "sidereal day" is the time it takes the Earth to rotate exactly once, which is not quite the same as a 'calendar' day.

The remaining additional velocity we need to impart we term δv_{leo}:

$$v_{leo} = v_{eq} + \delta v_{leo}$$

Isaac Newton had a rather appealing way of thinking about orbit and velocity. In his notes, there is a hand-drawn picture of a canon. As the power of the canon increases, it becomes clear that the canon-ball will eventually fall beyond the horizon, since the Earth's surface is curved. Newton realised that if the ball left the canon's muzzle with sufficient velocity, it would *never* fall back to the ground.

It would instead become a satellite of Earth, endlessly orbiting.

With the Law of Gravity he had discovered, and the understanding of 'Mechanics' he had pioneered, he could calculate the precise form of the orbit, which would be elliptical. If you wanted a circular orbit, he could also calculate the 'elevation' (angle above horizontal) the canon would need.

Of course, no one knew how to make a canon that powerful.

And neither man nor spacecraft could withstand the acceleration.[4]

Were Newton to continue improving his canon, the ball would enter an ever higher orbit. More of its initial kinetic energy would go into attaining altitude, and less into orbital motion. The equation opposite relating v to r, tells us that it moves more slowly around a higher orbit (as the square root of r). Eventually, there must come a point where the ball will remain synchronised with the Earth's rotation and fixed over some point on the Earth's surface directly below. We can calculate the orbital radius r_{geo} and thus average altitude h_{geo} of such a 'geostationary' satellite:

$$r_{geo}\,\omega^2 = \frac{GM_E}{r_{geo}^2}$$

$$\therefore \quad r_{geo} = \sqrt[3]{\frac{GM_E}{\omega^2}} = \sqrt[3]{\frac{6.674\times10^{-11}\times5.972\times10^{24}\times\left(8.616\times10^4\right)^2}{4\pi^2}}\ \text{m} = 42{,}162\ \text{km}$$

$$\therefore \quad h_{geo} = 42{,}162 - 6{,}378 = 35{,}784\ \text{km}$$

Note that, once again, the mass of the satellite is immaterial.

We can easily compute the orbital velocity:

$$v_{geo} = \sqrt{\frac{GM_E}{r_{geo}}} = \sqrt{\frac{6.674\times10^{-11}\times5.972\times10^{24}}{4.216\times10^7}} \simeq 3\ \text{km/s}$$

which is not daunting, unlike achieving the necessary altitude.

4 A typical rifle or canon imparts all the acceleration in a fraction of a second. If we assume our giant canon does so over one full second, low Earth orbit (LEO) would require an acceleration of 10 km/s², *i.e.* around 1,000 g.

To calculate the energy and initial velocity required, we can no longer use the simple equations applied earlier for low Earth orbit (LEO) as g varies significantly once height is comparable with R_E.

We need to consider what's happening in a bit more detail.

We must understand that the energy we assign to any body which is not accelerating is purely a *relative* measure, not an absolute one. A spaceship poodling around between the stars, and nowhere near any of them, might (arbitrarily) be assigned *zero* energy. Should it then drift near a star, its velocity — again a relative measure — will decide what happens. If its kinetic energy is sufficiently high, and it doesn't pass too close, its path will be deflected but it will remain free to continue its journey. There is a critical (relative) velocity above which our craft will always escape.

Below the "escape velocity" V_e, it will be captured, and enter an elliptical orbit.

A spaceship at rest on the surface of the Earth must perform work against the force of gravity to attain altitude. In other words, it must expend energy somehow, whether it rises in an elevator, is fired from a canon or invokes a wonderful new antigravity motor. Since we have assigned zero as the energy it would have if it entirely escaped our world altogether, we must further assign it *negative* energy while it is captured, which might seem rather strange, until we remember that all is relative.

We could equally assign zero as the energy it has on the Earth's surface and thus a positive value out in space, but we will then find ourselves returning to a negative value should we land on some more massive planet. The only truly general state we can choose for zero is freedom from gravity.

The same holds for the energy of an electron subject to the attraction of an atomic nucleus. A free electron is assigned zero energy, a captured one negative energy. Just how negative depends upon the atom concerned and how any existing electrons there are configured. This simple model can explain all chemistry. (You need only be concerned with a single electron in most chemical reactions.)

The force holding our spaceship (or electron) captive is called a 'potential'. An every day situation that is similar is a bucket of water at the bottom of a well. Much work is needed to haul it up. Hence, it is common to refer to any such mechanism as a "potential well".

Unfortunately, we live at the bottom of a very deep potential well. Escaping, in order to reach the almost infinite riches beyond, may well prove to be a matter of life or death for our civilisation, if not our species, which might continue in its traditional manner, fighting over whatever resources remain.

Recall Newton's Law of Gravity, and the fact that work done equals force × distance. Because the force varies with distance, we cannot simply multiply two magnitudes. Instead, we break up our ascent into many small steps of equal height δr and sum the energy required to lift one kilogramme:[5]

$$E_h^1 \simeq \sum_{r=R_E}^{R_E+h} \frac{G M_E}{r^2} . \delta r$$

5 The Greek letter 'Σ' (a capital 's') denotes 'sum'.

Now we make use of a marvellous trick. We choose a step height δr small enough so that we can ignore any error due to its size — in particular, any contribution made by the square or higher power. (Remember, anything less than unity grows smaller as you raise the power, not greater.) We then refer to the step height as 'dr'. A method known as the 'calculus', pioneered by Isaac Newton and Gottfried Liebnitz in the 17th Century, may then be used to determine the sum, or 'integral': [6]

$$E_h^1 = \int_{r=R_E}^{R_E+h} \frac{G M_E}{r^2}.\mathrm{dr} = G M_E \int_{r=R_E}^{R_E+h} \frac{1}{r^2}.\mathrm{dr} = g R_E^2 \left[-\frac{1}{r} \right]_{R_E}^{R_E+h} = g R_E^2 \left(\left[\frac{1}{R_E} \right] - \left[\frac{1}{R_E + h} \right] \right)$$

To escape from gravity completely, we would need $h \to \infty$:

$$E_e^1 = g R_E = 62.6\,\mathrm{MJ} \qquad \Rightarrow \qquad V_e = \sqrt{2 g R_E} = 11.2\,\mathrm{km/s}$$

By expressing geostationary altitude h_{geo} as a multiple of R_E, we can deduce:

$$E_{geo}^1 = 0.849\, E_e^1 \qquad \Rightarrow V_{geo} = 0.921\, V_e$$

In other words, it requires 92% of escape velocity just to reach geostationary *altitude*. To this must be added the 3 km/s orbital velocity. (Remember, the rotational velocity of Earth can help a bit.)

Arthur C. Clarke pointed out the use of geostationary orbit for communication satellites ('comsats') in the 1940s. (For this reason, it is sometimes called "Clarke Orbit".) A ring of just three can relay signals between any two points on the surface of the Earth. Though comsats are less massive now, the sheer expense of getting them into geostationary orbit has given rise to an alternative solution. Satellite telecommunication can instead use many more comsats, but in far cheaper LEO. Data packets can be sent via an optimal path between vehicles, computed on the fly. This also affords redundancy. Should a single geostationary comsat fail, half the planet would be out of touch. With a LEO system, all that is required is adaptive routing. Any failed satellite would also be far cheaper to replace or repair.

But we may have another use yet for geostationary orbit.

Rockets

So, to get into orbit, or to escape the Earth's gravity altogether, we need to accelerate to the right speed somehow. The speed necessary is measured in *kilometres*, not metres, per second. Newton imagined the use of a cannon to achieve this, as did the early science fiction author Jules Verne.[7] However, this

6 The name 'calculus', like the term 'calculate', derives from the Latin for 'stone', as used for counting. Once we have reduced δr to dr, we exchange Σ for the 'integral' sign $\int$, which also denotes a sum, but of infinitessimal amounts. See the Appendix for an introduction to the calculus, and a brief (but adequate) introduction to integration.

7 *De la Terre à la Lune* (From the Earth to the Moon), 1865.

highlights the same problem we encountered in the design of a roller coaster: that a human can only withstand perhaps 6 g, for a few seconds only, and then only if they are extremely fit and healthy.

Verne was to inspire one Konstantin Tsiolokovsky to propose something better.

We have already met Newton's Second and Third Laws:

Force = mass × acceleration

In equilibrium, every force is met with an equal and opposite reaction.

The above statement of the Second Law is a little imprecise for our purposes. A new concept must be introduced: the 'momentum' of a body expresses how hard it is to bring to rest, and is obtained by multiplying mass and velocity. A more general statement of the Second Law is:

Force = rate of change in momentum

which allows velocity to remain constant and mass to vary instead:

Force = velocity × rate of change in mass

Imagine yourself standing on a stationary open railway truck, with a pile of rocks behind you. You throw a rock out the back. If the wheel bearings are well greased, the truck should move in the opposite direction. The greater the speed you give the rock, the faster the truck will move the other way.

This is an example of the *conservation of momentum*: the momentum of the rock is balanced by that of the truck. The momentum of the rock can be considered negative; that of the truck, positive.

If you keep throwing rocks out behind, the truck will continue to accelerate. Better still, because the mass of rock, and thus that of the loaded truck (m), is diminishing (by δm), the velocity change with each rock will gradually increase. If only we could throw rocks without carrying them with us!

Now, we have an illustration of Newton's Third Law: the product of the rock's mass and its velocity is equivalent to a force, which will be met with an *equal and opposite reaction*. That reaction will impart an acceleration δv_t to the truck, according to its mass:

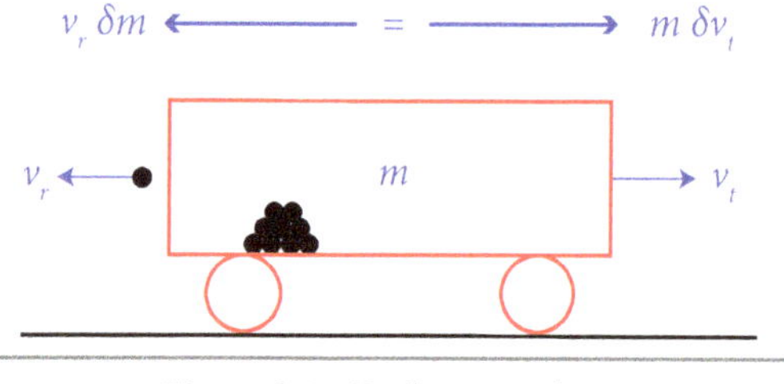

Figure 3.1 Rocket propulsion.

$$m\,\delta v_t = -v_r\,\delta m \qquad v_r = \text{velocity of rock}$$

This is why a balloon whizzes around the room when you blow it up and release it. Any machine which expels material in one direction in order to propel itself in the other is called a 'rocket'.

Tsiolkovsky observed that a rocket would work even in space, where there is nothing to push against, and that it could lift people up there without crushing them.

He also "did the maths", to account for a continuous flow, rather than a lump at a time.

Again, we can employ an infinitessimal and use the calculus:

$$\Delta v_s = -v_{ex}\int_{m_1}^{m_2}\frac{1}{m}\,\mathrm{dm} = -v_{ex}\left[\ln m\right]_{m_1}^{m_2} = -v_{ex}\left[\ln m_2 - \ln m_1\right]$$

m_1 = mass of vehicle at start
m_2 = mass of vehicle at end

$$\therefore\ \Delta v_s = v_{ex}\ln\frac{m_1}{m_2} = v_{ex}\ln M_r$$

Tsiolkovsky Rocket Equation

This is one form of Tsiolkovsky's "rocket equation", relating the change in velocity of a spaceship Δv_s when a mass $(m_2 - m_1)$ of 'propellant' is emitted with (constant) exhaust velocity v_{ex}.

The ratio of initial to final mass is known as the "mass ratio" M_r. Clearly, the bigger the mass ratio, the bigger the "delta v", but unfortunately there are limits to just how light we can make a spacecraft, and hence to the value of M_r which can be achieved, according to payload and technology. If we could make a lighter spaceship then we could increase Δv_s with the same quantity of fuel. It is the mass ratio which defines the inefficiency of a rocket. (The same may be said of a car, which typically weighs over a tonne, yet carries only about 1.2 passengers, on average, who weigh less than a tenth as much.)

With chemical fuel, propellent exhaust velocity v_{ex} is limited to about 4.5 km/s. Nuclear reactions can certainly improve on this, but come with their own problems. While experiments were undertaken in the 1950s, nuclear fission remains an unlikely candidate since the fuel is, by definition, extremely heavy and is *not* ejected; it must be carried *in addition to* the propellent (decreasing M_r, and thus Δv_s). Environmental concerns should also rule it out. A better option, developed in Britain in the 1970s, is the "ion thruster", where charged particles are accelerated by an electric or magnetic field. Ion thrusters are indeed already in use for manœuvering small unmanned spacecraft, but are limited to such rôles because of their relatively low power. While v_{ex} is very high (around ten times that of a chemical rocket), the rate of mass ejection is self-limiting. The idea also only works in space.

A far better but much more distant prospect is nuclear *fusion*. A design for an interstellar spacecraft in 1978 proposed pulsed detonation by high energy lasers.[8] Such a vessel would have to be constructed in space, and would require significant technological advance, but remains plausible for the future.

The most practical fuel, used by both the Saturn V and Space Shuttle, is hydrogen. Both fuel and oxidant must be liquefied for compactness. The liquid oxygen comprises eight ninths of the mass.[9] Burning hydrogen and oxygen liberates only steam, which eliminates the worst environmental concerns on the Earth's surface. v_{ex} is competitive at 3.6 km/s. If $M_r = e$ then $\Delta v_s = v_{ex}$ — only a third of Earth's escape velocity. To escape the Earth, we would need a mass ratio of e^3 (≈ 20), which is quite impracticable. Given known fuels and structural materials, the limit is closer to e^2 (≈ 10).

8 *Project Daedalus*. 1978. Martin, A. R. (Ed.), Supplement, Journal of the British Interplanetary Society.

9 Oxygen has molecular mass sixteen and hydrogen two, giving a total of eighteen for one of each, as required.

The truth is unfortunately worse than this since we have accounted for neither gravity, reducing the rocket's effective acceleration, nor air resistance. These are *secondary* effects: rockets typically offer an acceleration several times greater than gravity, and can possess a very small cross-section.

We are unable to (yet) design an ordinary rocket that can even reach orbit in a single bound.

Making the rocket bigger can increase the payload — keeping M_r constant — but not Δv_s.

Tsiolkovsky understood this perfectly, and offered a solution: make the payload itself another rocket. One can do this recursively: the payload of the second rocket can be a third one *etc.* With such a 'multi-stage' (or 'step-') rocket, Δv_s can be multiplied by the number of stages (steps).

While Δv_s scales linearly with the number of stages, M_r scales exponentially, leading quickly to vehicles of enormous total mass. If it grows too big, the vehicle will be crushed under its own weight. The limit is now set by fuel density and the strength-to-weight ratio of available structural materials.

Again, the truth is actually worse, because M_r does not properly account for the scaling of mass with staging. In our calculation, the final mass of the rocket m_2 included not just the payload but the structure of the rocket itself. The *payload ratio* may be significantly greater. For example, suppose the mass ratio of each stage were 5 : 1. Should the structure mass four times the payload, the payload ratio would be 20 : 1. The final payload ratio of a three-stage vehicle would then be *8,000* : 1. Fortunately, things are nothing like this bad, mainly because of the enormous power of the first and second stage. Their payload includes the remainder of the vehicle, as well as the final payload itself.

The Saturn V was a three-stage rocket with a final payload ratio nearer 50 : 1, which was quite an achievement. Saturn managed to launch some 60 tonnes (the complete Apollo spacecraft) with Earth-escape velocity, yet possessed a fuelled mass on the ground of only about 3,000 tonnes. That represents something close to the limit of what is technically possible, even today.

So far, we have only spoken about the cost of leaving the Earth. Sadly, this is only half the problem in visiting another world. Our craft must also decelerate upon arrival, and somehow return. Propellent had be carried to the Moon to 'inject' the Apollo spacecraft into a lunar orbit. It would otherwise have simply flown past, as it's relative velocity was significantly greater than that for lunar escape. Even more was required to allow a controlled descent, and avoid crashing on the surface. Missions to Mars have employed aerobraking in the Martian atmosphere, burning away a "heat shield".

Returning home requires achieving the local escape velocity. Fortunately, lunar gravity is much weaker than the Earth's, allowing Apollo to employ a single-stage ascent. Mars is a tougher call.

If we assume, for simplicity, that the world we wish to visit has the same mass as Earth, and thus the same escape velocity, then we need our rocket to provide twice the "delta v" it would for a one-way trip. As can be seen by inverting the Tsiolkovsky Rocket Equation, this has a disastrous consequence for the mass ratio of the required rocket:

$$M_r = e^{\frac{\Delta v_s}{v_{ex}}}$$

$$\therefore \quad M_r^1 = e^{\frac{V_e}{V_{ex}}}, \qquad M_r^2 = e^{\frac{2V_e}{V_{ex}}} = \left(M_r^1\right)^2$$

The mass ratio needed for a rocket carrying enough propellent to escape *both* gravity 'wells' would be the *square* of that needed for a one-way trip. If a thousand tonnes of propellent were required for a craft to escape the Earth, a *million* tonnes will be necessary for a return trip, which is quite impossible.

A Mars return mission will require acquiring propellent locally.

An *interstellar* mission is much tougher still. Tsiolkovsky's simple innocent equation harbours a terrible truth: that we are prisoners in our cradle — our solar system, if not our planet. It tells us there is no possibility of carrying enough chemical propellent for deceleration on arrival. You cannot aerobrake in a stellar atmosphere; even if available technology permitted adequate radiation shielding, the additional mass it would impose would be self-defeating. The thermonuclear system of *Project Daedalus* may offer a solution, but is likely to remain beyond our technological (and economic) means for generations, and would, in any case, first require the total mastery of interplanetary flight.

Perhaps, this barrier to interstellar flight is for the best, since it requires us to progress socially as much as technologically. No one nation could succeed in such an undertaking. Only the result of considerable human integration could accomplish the development and construction required. It can equally be argued that only such a project could ever succeed in *bringing about* such integration.

Almost all our current technology originated in war, or the preparation for it. Perhaps, in space exploration, we might find an alternative, and ultimately more productive, challenge.

Space planes

Very little can be accomplished as long as we remain dependent upon huge multi-stage rockets which are used just once then thrown away. When the vehicle required to launch a 60-tonne spacecraft weighs three *thousand* tonnes, the exploration and exploitation of space will be prohibitively expensive.

Manned spaceflight, in particular, can make little progress.

What we need is a machine that offers a reusable "single stage to orbit" (SSTO).

The original design for the Space Shuttle was a "two stage to orbit" (TSTO) solution, where 'booster' and 'orbiter' were each fully reusable and possessed wings and both rocket and jet engines. The scale, complexity and development cost of the booster eventually ruled it out. In turn, the complexity of the orbiter made it as expensive to refurbish and launch as a throwaway rocket.

A new and radical approach is offered by the British SABRE engine — the brainchild of Alan Bond, who was also lead investigator for *Project Daedalus*. SABRE removes the need to carry oxidant for the ascent through the atmosphere. It has long been understood that one way of achieving a SSTO is to fly as a conventional aeroplane while in the atmosphere. The problem has been (as the designers of the

Space Shuttle discovered) that the weight of *additional* jet engines receives inadequate compensation through the saving in oxidant. SABRE combines jet and rocket into a single hybrid engine.

It also embodies all that has been learned about efficient high performance jet design.

One essential trick is to cool incoming air, which would otherwise be hot enough to destroy all but the most massive intake. Extremely innovative design has produced an ultra-lightweight intake capable of cooling air from in excess of 1,000 to around −150 °C in just ten milliseconds. Supercooling incoming air also raises the rate at which mass passes through the engine, increasing thrust.

Conventional jet thrust is used for the first 25 kms of the ascent to orbit, burning hydrogen with atmospheric oxygen. The intake is then sealed and the engine transforms into a rocket, again burning hydrogen, but now with stored liquid oxygen. Because oxygen is eight times heavier than hydrogen, a significant proportion of the initial mass is reduced by eight ninths. Environmental impact is minimised, with 98% of nitrogen oxides removed from the exhaust, leaving little but water vapour.

SABRE is to power a SSTO vehicle called *Skylon*, which is intended to be fully reusable. It is expected to reduce the cost of getting a kilogramme into orbit from around $20,000 to perhaps $1,000.

3.3 An elevator into space

How high can a tower be?

Space planes will form an essential step in the evolution of orbital launch vehicles but fully uninhibited exploration and exploitation requires an even greater reduction in cost, down to hundreds, perhaps even tens, of dollars per kilo.

We began this chapter with a very simple calculation of the energy required to raise a kilogramme above the atmosphere (1 MJ). That required to get a kilogramme to Earth-escape velocity was then computed. It came to just 63 MJ, or about 17 kWh, which costs around £2.50, how ever you buy it.

Clearly, we're doing something wrong when we spend even the $1,000 with Skylon.

It fell to a young Russian engineer, one Yuri Artsutanov, to suggest the ultimate solution, in 1959. He may have been inspired by Tsiolkovsky, who was in turn inspired by the sight of the (then new) Parisian triumph of Monsieur Eiffel, to suggest building a tower all the way up into space.

That ain't gonna happen.

The first thing we might consider is a cylindrical tower. What we will need is the material with the highest available *compressive* strength, so that the base will not be crushed under the weight of the rest, and the lowest density, so that the tower will be as light as possible. We must be grateful for the Romans, who bequeathed us concrete. The very best concrete can (in theory) withstand 70 MPa (million newtons per square metre), and can achieve a density ρ of around 2,250 kg/m^3.

The weight of the column must never exceed the compressive strength of the base:

$\rho g h A \leq \sigma A \qquad \beta = \frac{\sigma}{\rho g}$

h = height, A = area of cross section
σ = compressive strength of material

$\therefore\ h_{\max} = \beta \qquad h^c_{\max} = \beta^c \simeq 3{,}170\,\mathrm{m}$

β thus represents a "figure of merit" for tower-building materials.

By comparison, stainless steel has a much higher density, of about 8,000 kg/m^3, but also a much higher compressive strength, of up to 310 MPa, imparting a slightly greater maximum column height:

$$h^s_{\max} = \beta^s \simeq 3{,}950\,\mathrm{m}$$

Steel is now the material of choice for the construction of vertical buildings, though we don't yet approach the maximum height technically possible. Three considerations limit what we attempt. First, those who abide within and below will demand a safety margin. Second, the actual materials used cannot be expected to perform as well as the best sample in the lab. Construction methods will also weaken them — components welded or bolted together will never be as strong as a homogeneous solid. Lastly, allowance must be made for flexing in the wind, and possibly earthquakes.

Every component and join must withstand the *peak* of a varying load.

The design of any structure is a complicated business, keeping many engineers employed. Here, we shall interest ourselves only in the upper bound to what is possible.

Of course, a solid building is only rarely of any use. It might be suitable for mounting a statue or an antenna, but it can accommodate neither office nor apartment. However, if we hollow out our column we find that nothing changes. All we bring about is a reduction in the effective area A, which appears on *both* sides of our starting inequality above. Both the weight of the column and the ability of the base to support it are reduced in equal measure. β remains the same, for any given material.

Aside from allowing us to use the space within, hollowing out a structure reduces its cost dramatically, which, to a large degree, depends linearly upon the total volume of material used.

Clearly, a simple column is not going to get us into space; at least, not one made with steel or concrete. It is also obvious that a tall and slim building will not easily withstand the wind. While steel will survive a lot of bending and twisting, something nearly four kilometres high is a "big ask".

We need something a lot higher and a lot less flexible.

The solution is to broaden the structure in proportion to the weight above each point. The total strength of each slice will diminish as we ascend, in proportion with the load relieved (Figure 3.2):

$$\sigma \delta A = -\rho g A(h) \delta h$$

Taking the limit, as $\delta h \to 0$ and integrating:

$$\int_{A_0}^{A_h} \frac{1}{A} . dA = -\frac{h}{\beta} \qquad \therefore \quad \ln A_0 - \ln A_h = \ln \frac{A_0}{A_h} = \frac{h}{\beta}$$

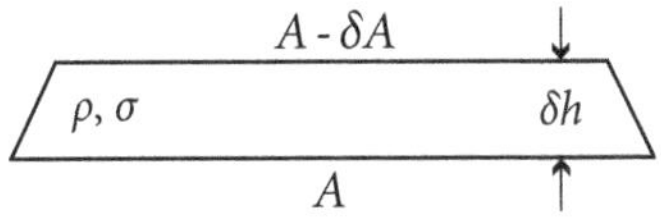

Figure 3.2 Segment of tower.

$$\therefore \quad A_0 = A_h\, e^{\frac{h}{\beta}}$$

In theory, we could build a tower as high as we like, just as long as we can find an area A_0 for the base… which grows by a factor e for every β in height. Its cross-section varies exponentially.

Suppose we need a platform at the apex of area 100 m^2 (*i.e.* 10 × 10 m), and a height, to rise above the atmosphere, of 100 km, then:

$$A_0 \simeq 100\, e^{25} \simeq 7.2 \times 10^{12}\ \mathrm{m}^2$$

If the base were circular, it would have a radius of more than a thousand kilometres.

The problems involved in constructing something the size of a continent would no doubt be 'challenging', but the cost would obviously be prohibitive, even for a united world. (The cost, recall, is proportional to the volume, which could be recovered by integrating $A(h)$, with respect to h.)

A much better idea is to build the tower inverted, with the pointy end on the ground.

Application

3.1 Calculate the base radius required for a tower of height 10 km, supporting an hotel with area 10,000 m^2 and constructed from steel (density ~8,000 kg/m^3 and compression strength ~240 MPa).

Given a steel at \$50 per tonne, and a structure using just 1% of the volume, estimate the cost.

A tower down from space

This is what Yuri Artsutanov proposed: start in geosynchronous orbit and build a tower *downwards*. It must have been one of those occasions when jaws drop as the laughter dies down. The more thought was applied, the more objections dissolved. Perhaps every genius begins with derision.

Of course, he had to start in geosynchronous orbit for the base to remain above the same point on the ground. But that is at an altitude of ~36,000 km — two orders of magnitude higher than our previous tower. We must be more careful in our calculations. Because the height is large when compared with the Earth's radius R_E, we must allow for continuous variation in gravity. You might think that will help us, since the massive base will reside where gravity is weakest, but you'd be wrong. The integration is of the same function, and between the same bounds, whether we build up or down.

However, there is another effect which does allow us to benefit from both altitude and direction. We must also account for *inertia*.

What we would build is equivalent to a gigantic slingshot, rotated by the planet to which it's attached (Figure 3.3). Each segment of our tower will be doing its best to keep travelling in a straight line. We will perceive this as an acceleration upwards, counteracting gravity. There is, in reality, no such acceleration. The truth is that some of the gravitational acceleration downward is "used up" keeping the segment where we want it to be, at a constant altitude instead of flying off at a tangent.

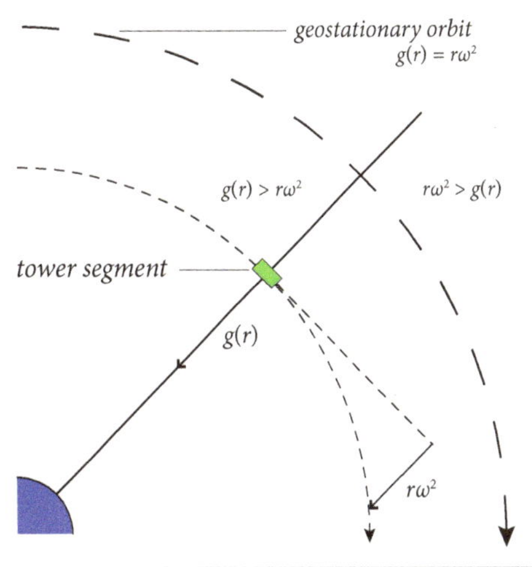

Figure 3.3 Acceleration of tower segment.

Remember we are forcing (using our slingshot) each segment to move slower than it would were it in orbit. The inertial 'acceleration' is less, often much less, than it would be then. But the overall effect on the entire tower is to dramatically reduce the necessary volume of material, and hence its cost.

This time we need a material with a high *tensile*, not compressive, strength. It's worth taking a little time to consider the difference between the two, and where tensile strength originates.

Compressive strength is always limited because of the tendency of even a pure crystalline or metallic material to shear. Curiously, defects in the lattice tend to *prevent* shear, and thus improve performance. Tensile strength is usually much greater because shear is less of an issue. Tensile strength is purely down to that of the chemical (metallic, ionic or covalent) bond between atoms. Defects, usually due to impurities, will *reduce* performance because they offer an opportunity for fracture to begin.

Only outermost electrons ever take part in a chemical bond. All others are of no account, and the nucleus just adds unwanted mass, and hence density. To gain the highest β, we need the lightest possible material, with the lowest atomic mass, and the strongest chemical bond. Again, someone thought they were joking when they suggested solid hydrogen, as the lightest element with an extremely strong (metallic) bond. Then they read of experiments concerning that material. Solid hydrogen might be expensive to manufacture, but the gaseous kind is relatively cheap and we'll have to build dedicated factories in space, whatever we use, easing solidification. The big problem here is something else.

In an atmosphere with lots of oxygen, solid hydrogen makes one hell of an explosive.

Moving along the Periodic Table, looking for something light, cheap and safe, we come across carbon. No one knows the tensile strength of diamond, but it's at least 60 GPa (*i.e.* 60,000 MPa), and theory suggests it may be over 200 GPa. It's so high that measurement has thus far proved impossible.

Its density is rather high, at around 3,500 kg/m^3, but that still leaves a β of somewhere around five *thousand* kilometres. Steel, in comparison, can manage one of just twenty-five.

Manufacturing a gigatonne of pure diamond is currently just a tad expensive. While melting diamond is not too hard, welding components together *perfectly* is an art we do not yet possess. Yet there has been progress: in 2014, carbon 'nanothreads' were successfully made, from common benzene. Their structure, and thus their tensile strength, is similar to that of diamond.

Carbon atoms can be woven into a nanoscopic cylinder of indefinite length. So called carbon 'nanotubes' posses all the tensile strength of diamond — up to around 300 GPa, in theory — but only a fraction of their density, at about 550 kg/m^3, conferring a β of around *fifty-six* thousand kilometres.

As they say, this oughta do it.

Before we tackle Artsutanov's inverted tower down from geosynchronous orbit, we might ask whether a simple cable might hang from a satellite, all the way down to the ground. This would demand sufficient tensile strength σ to sustain the cable's weight at the base, after allowing for inertia:

$$\sigma A \geq \rho A \int_{R_s}^{R_o} \left(g(r) - r\omega^2 \right).dr \qquad g(r) = g\frac{R_s^2}{r^2}$$

$$\therefore \quad \beta \geq R_s^2 \int_{R_s}^{R_o} \frac{1}{r^2}.dr - \frac{\omega^2}{g} \int_{R_s}^{R_o} r.dr \quad = R_s^2 \left| -\frac{1}{r} \right|_{R_s}^{R_o} - \frac{\omega^2}{2g} \left| r^2 \right|_{R_s}^{R_o}$$

$$\therefore \quad \beta \geq R_s^2 \left(\frac{1}{R_s} - \frac{1}{\alpha R_s} \right) - \frac{\omega^2}{2g} \left(\alpha^2 R_s^2 - R_s^2 \right) \qquad \alpha = \frac{R_o}{R_s}$$

$$\therefore \quad \beta \geq R_s \left[\frac{\alpha - 1}{\alpha} - \frac{\gamma}{2} \left(\alpha^2 - 1 \right) \right] = R_s \psi(\alpha) \qquad \gamma = \frac{R_s \omega^2}{g}$$

The relation holds for any value of α, not just that for r_{geo}. (γ is a planetary constant, representing the ratio of inertial to gravitational acceleration experienced by an object on its surface.)

It points to a rather startling result:

$$\therefore \quad \beta_E \geq 0.7749 R_E \quad \simeq 4{,}940\text{km} \qquad (\alpha_{geo} = 6.611 \quad \gamma_E = 3.458 \times 10^{-3})$$

If we can find a material which can suspend a cable of just 5,000 kms under surface gravity (1g) then it could suspend the entire *36,000* kms from geosynchronous orbit. Arthur Clarke called this rather critical figure the "escape length" for Earth, comparable with the escape velocity for a rocket.

A simple cable is, however, more than a bit wasteful. We will not need the same area of cross section half way down, to suspend half the cable, as at the base, to sustain all of it. If we employ an exponentially tapered tower then only a fraction of the volume will be needed, at a fraction of the cost.

And the cost will be pretty high, per unit volume, given that pure diamond is the only material we know how to manufacture in quantity. In truth, a space elevator will only be plausible if we can learn how to manufacture continuous interwoven carbon nanotubes or nanothreads.

So we must return to the mathematics of a tower, but this time account for the variation in both gravity and inertia with altitude:

$$\sigma\delta A = \rho A\left(g\frac{R_s^2}{r^2} - r\omega^2\right)\delta r$$

$$\therefore \quad \beta\int_{A_o}^{A_s}\frac{1}{A}.dA = R_s^2\int_{R_o}^{R_s}\frac{1}{r^2}.dr - \frac{\omega^2}{g}\int_{R_o}^{R_s} r.dr$$

We have performed this integration before, but with the limits reversed. If we note that:

$$\ln\frac{X}{Y} = -\ln\frac{Y}{X} \qquad \text{and} \qquad \int_a^b f(x).dx = -\int_b^a f(x).dx$$

then we have:

$$\beta\ln\frac{A_o}{A_s} = R_s\,\psi(\alpha) \qquad \text{where} \qquad \psi(\alpha) = \frac{\alpha-1}{\alpha} - \frac{\gamma}{2}\left(\alpha^2-1\right)$$

$$\therefore \quad \underline{A_o = A_s e^{\frac{R_s\psi(\alpha)}{\beta}}}$$

If the base of the tower coincides with geosynchronous orbit then α may be expressed in terms of γ — the constant ratio of inertial to gravitational acceleration on the planetary surface:

$$r_o\omega^2 = g\frac{R_s^2}{R_o^2} \qquad \Rightarrow \qquad \alpha^3 = \frac{1}{\gamma}$$

$$\therefore \quad \psi(\alpha_{geo}) = 1 - \frac{3}{2}\sqrt[3]{\gamma} + \frac{1}{2}\gamma \qquad \Rightarrow \qquad \psi_E(\alpha_{geo}) = 0.775$$

Now we can see what kind of "real estate" will be needed in geosynchronous orbit.

	σ (GPa)	ρ (kg/m³)	β (km)
steel	2	8,000	25
carbon fibre	4.1	1,750	240
Kevlar	6.4	1,800	360
diamond	~150	3,520	~4,500
carbon nanotube	~300	550	~55,600

First, let's see how well we can do without cheap diamond or carbon nanotube.

The Earth station is likely to be busy, so let's assume it's big; say, a square kilometre. A Kevlar space elevator would need to occupy nearly a *million* square kilometres in space. If the orbital station were circular, it would have a radius of around five hundred and fifty kilometres.

While this seems extreme, we must remember that such a project might engage the economic and technical co-operation of the entire human civilisation. Should we successfully develop the means to manufacture and handle carbon nanotubes or nanothreads, the orbital station shrinks to just over one square kilometre — in other words, we no longer need a *tower* at all; just a cable, or cables.

Even a diamond tower would require a base just three times the area of its Earth station.

There's one more consideration: we will need to keep the *centre of gravity* in geosynchronous orbit, and our tower will mass gigatonnes. Given a tower, where most of the mass is concentrated at the base, we might simply start higher, so as to leave the centre of gravity where we need it. But such a tower would be much more massive and thus much more costly. Anyway, for the project to become remotely affordable, something like the carbon nanotubes/nanothreads are essential, and obviate a tower.

Just two possibilities remain.

We might construct an equal and opposite tower *upwards*, above a geosynchronous station. Strange things happen when we continue outward. The inertial effect will exceed gravity, keeping the new tower also in tension and making our elevator 'fall' upward. This would be marvellous for the launch and recovery of spacecraft. Unfortunately, such an upward tower would have to be much taller, since it truly is the centre of *gravity* that matters, not the centre of mass, and gravity grows weaker with the distance from Earth. Hence, this approach more than doubles the cost.

Though it sounds like lunacy, there is a better option: nudge an asteroid into Earth orbit.

Obviously, it would have to be (initially) at least as massive as the tower it must balance. A sufficiently large carbonaceous asteroid could also provide the raw material for its construction.

Asteroids have very large orbits, which allows a very long lever. Hence, the deflection need only be minute, but would also have to be extremely precise. An error could cause impact on Earth.

In fact, this is a capability which we have no option but to develop. Asteroids, and comets, are deflected anyway by mutual collision and the gravitational influence of planets. Given their relative velocity, and thus kinetic energy, even a small object can cause catastrophe. In 1908, an object detonated above a forest in a remote area in Siberia, called Tunguska. Although estimated at just 50 m wide, it levelled around 1,000 square kilometres — similar in scale to the impact of a nuclear fission bomb.

The impact of an object 1 km across might extinguish our species.

In case you felt inspired by one Hollywood movie or another, blowing up the offender is not a good idea. That just creates a host of smaller (possibly radioactive) threats.

Currently, all we can do is to locate and monitor as many objects as possible which have trajectories which cross the Earth's orbit. The joint ESA/NASA Asteroid Impact and Deflection Assessment (AIDA) represents a first step in learning how to actually respond to a serious threat.

Once we have perfected the art of deflecting asteroids, the risk of deliberately capturing one above our heads should prove acceptable. After all, the same risk applies to managing one already set to collide. Capture can be carried out in steps, to minimise, or even perhaps eliminate, the risk.

Meanwhile, the benefit may be to dramatically increase our prosperity.

Even vacations in space may become affordable. Someone has already calculated just what a hotel in geosynchronous orbit would be worth, but the best locations might be along the tower itself.

Just imagine the view from, say, a hundred kilometres up.

It should not surprise anyone that many technical difficulties exist with both the operation and the construction of an orbital tower and space elevator. One of the paramount operational problems lies in the design of the trains. They need to operate at very great speed since the journey is almost equivalent to circumnavigating the planet. Electromagnetic propulsion, as well as suspension would seem the only choice. The difficulty is in getting energy to them. Losses along a 36,000 km conventional cable would be unacceptable. Superconducting cables would be nice, and rendered easier to accomplish in the cold of space, but they are likely to remain far too expensive to span the tower height.

Locating nuclear reactors within track or train track would prove unpopular with folk below.

We might 'beam' power to the train. One plan calls for the use of high-energy lasers, targeting photovoltaic cells. Pointing what would otherwise form directed-energy weapons at carriages with people inside seems an unwelcome prospect, especially given an efficiency probably less than 1%.

We should stop to think just how much energy is needed.

Very little energy is required by an elevator because carriages can counterbalance each other. The gravitational potential energy of a falling kilogramme precisely compensates that of a rising one. We can eliminate friction by using electromagnetic suspension of each carriage, as with a 'MagLev' train, whose propulsion system can also serve as a 'regenerative' braking one, with very little loss.

Provided falling and rising traffic closely match, very little energy is required for operation.

Our problem of power transmission is also resolved.

There is no need to transmit power along the structure, only *across* it.

One more matter must be accounted for: *rotational* energy.

Each section of the tower is rotating around the Earth with the same *angular* velocity, but its (instantaneous) linear velocity, and thus its kinetic energy, will be higher, the greater the altitude. Rotational energy thus grows linearly with radial distance. As our carriage climbs, it acquires energy from the tower. Ultimately, the planet itself will supply this, via the slingshot effect. Its day will grow shorter, though the effect will be infinitessimal, given the mass involved. The problem is that a carriage moving upward will apply a bending moment to the tower, causing stress and deflection. Although the counterweight will act like a pendulum to correct the latter, there is a danger of oscillation.

If the transfer of rotational energy is not controlled, the structure might break.

Once again, we might appeal to the counteracting nature between traffic in each direction. If they closely match in number and mass then we might derive a schedule that would control deflection. Complex calculations would be needed for the timing of trains so as to minimize the amplitude of oscillation both of the tower as a whole and along its length. For just the same reason, soldiers are trained to break step when crossing a bridge. We must take great care to avoid *resonance.*

It's fascinating to note that such a project as this would be impossible without computers.

They would hardly be the first thing you'd notice.

Orbital ring

Just in case you thought an orbital tower 36,000 km downward from space seemed nuts, there's another (serious) proposal that is more 'loopy', in every sense.

We noted earlier that our tower or cable from geosynchronous orbit is long enough to circumnavigate the globe. Well, suppose we did just that instead. (The great Nikola Tesla actually suggested such a thing way back in 1919.) Each element of our ring would be in orbit. In order to keep it in tension, we might rotate it slightly faster than otherwise necessary.

In order to suspend a tower, we electromagnetically (to avoid any friction) suspend a station on the ring. To launch and retrieve spacecraft, we simply couple them to the ring for a brief interval.

By suspending multiple towers, like spokes in a wheel, we also gain an extremely rapid global transport system (which was Tesla's purpose, though he proposed a ring closer to the ground).

Look up at the stars at night. Maybe some race has already built one.

Application

3.2 Calculate γ, and thus ψ for a geosynchronous orbital tower on Mars.

3.3 Why would a (compression) tower up from the ground help with a (tensile) orbital tower down?

3.4 A "sky hook" can raise a payload into space from low Earth orbit (LEO). A cable, terminating with a hook, is rotated vertically around a spacecraft in LEO. If the rate of rotation ω_h is chosen carefully, the hook will appear to drop and then rise vertically from the Earth below. A payload may thus be attached, hoisted and given a substantial velocity via the slingshot effect.

Calculate the rotation rate ω_h necessary around a satellite in orbit at an altitude of 200 km.

Comment on any practical problems you can think of, and propose solutions.

Appendix: Four easy pieces

Counting beyond reason

The numbers we first learn about — 1, 2, 3, 4 *etc.* — can be termed 'natural' numbers, since they arise 'naturally' from *counting*, which also offers the simplest definition of 'number': a count. They also form one of the first examples of conscious *abstraction*. Even a quite small child can see that four apples and four oranges have something in common. (We hope that children everywhere experience fruit.)

Perhaps the earliest counting system was just a 'tally' stick, on which notches might be carved. This is fine while we're counting relatively few sheep (for example), but we need something better when our flock outgrows the longest stick we can find. (Putting notches ever closer together risks error.)

Notches are hard to erase, so traders later opted for a "counting board", placing pebbles in a tray of sand. This made adding and subtracting much easier, but didn't offer greater range.

It wasn't long before a common solution was found. You just count big gololłups alongside units. Roman numerals offer an example of this method, and survive to this day as an alternative enumeration where some distinction is useful. The preface of this book is paginated this way, to distinguish material about book and author from its actual content. Different symbols represent gololłups of different size: 'V' denotes five, 'X' ten, 'L' fifty, 'C' a hundred and 'M' a thousand. The system affords a compact notation for large numbers; for example, the year 2015 becomes MMXV.

But I wouldn't try performing much calculation. You will not find it easy.

Somewhere in northern India, a much better solution was discovered.

First you decide upon a particular number, called a 'base'. Then you can begin counting up from one. When you reach the base, you start a new column, which counts gololłups of that size. On exhausting that column, you begin another one, and so on. This is the system you learned at school, where the base was certainly ten, which is a shame because that is an awkward number. While you can share ten things easily between two or five friends, it's a bit of a problem otherwise. Which is why our more sensible ancestors commonly used twelve, which divides by two, three, four and six.

The important thing about this system is that it is the *position* of each digit that decides the size of gololłup it represents. Precisely the same notation may be used everywhere. All you need is a distinct symbol for each count from one up to one less than the base... and just one more.

Prior to the introduction of the positional representation of numbers, no one ever needed a symbol denoting *zero*. There can be little doubt that people tried merely leaving a gap where there was a column where no contribution was made. Equally, there can be little doubt that embarrassing errors arose in consequence. It will have soon been learned that a dedicated symbol for zero was required.

No one knows quite why a circle was chosen. Some believe that the symbols for other decimal digits began with the requisite number of included angles. (Try drawing them with only straight lines, and no curves.) A circle, with no included angles would then fit right in. Others believe the counting board explains it. On removing the last pebble, a small circle would remain. Maybe it was something else. What matters is that we all agree on the convention. It came with positional notation.

One other convention came along too: writing mathematics from right to left.

In Western Europe, we learned how to write text from the Romans, but mathematics from Arabic traders. Mathematics is typically written left-to-right, but is best read right-to-left.

After happily learning arithmetic, many students struggle with fractions. It often helps to reduce the problem to one of counting smaller things. For example, if we must refer to three fifths of a cake then we simply imagine cutting the cake into five (equal) slices and setting aside three of them. Adding two fifths to three fifths is no more challenging than adding two to three of anything. We can also convert an 'improper' fraction to a mixed number merely by asking how many whole cakes we can assemble and how many slices remain. This is why any fraction is called a 'rational' number.

It can still be understood simply through counting.

The Ancient Greeks believed they might explain everything they encountered in the world through counting. It came as a profound shock to find that counting was not enough.

After Pythagorus gave the world his famous theorem, relating the magnitudes of each of the three sides of a right-angled triangle, they discovered that the hypotenuse was sometimes *incommensurable* with one or both of the other sides. They were already concerned about the difference between counting a 'multitude' (of distinct things) and measuring a magnitude. They had also already solved the problem of finding the greatest common divisor of any two numbers, and naturally thought that any two distances could be measured by thus finding a stick of the right length.

They were wrong.

Some of the most profound changes in our understanding of the world have been the result of a master charging a student to determine something, only to have them return with a result which appears to be 'obviously' wrong. One can imagine that this was such an occasion.

An 'altercation' seems likely, with the student dispatched with a sore ear (or worse).

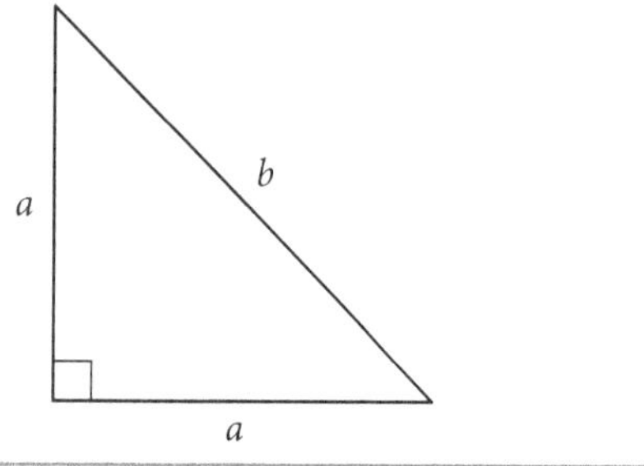

Figure 3.4 Right-angled isosceles triangle.

At some point the master would begin to wonder, and then take over the investigation (and perhaps later claim all the credit).

Arguably, the simplest example of incommensurability is in the simplest right-angled triangle: an isosceles one (Figure 3.4).

From Pythagorus' Theorem, we know that:

$$b^2 = 2a^2$$

Assume that a and b are integers, but share no common divisor. (Otherwise, we could use a longer stick.) This means that at least one must be odd. (Otherwise, we could use a stick twice as long.)

Only an even number has an even square. Hence, b must be even, so we can write:

$$b = 2c \qquad \therefore\ 4c^2 = 2a^2 \qquad \therefore\ a^2 = 2c^2$$

But this would mean that a is also even, and that a and b *do* share a common divisor — two — which contradicts our premise, that they do not. They cannot both be a simple count.

b/a cannot be expressed as a simple ratio, using two integers, and is therefore termed 'irrational'. As a consequence, it cannot be calculated exactly — precisely yes, but never exactly. Your calculator can give you as many digits as you like, but never the *exact* value.

π — the ratio between the circumference and diameter of a circle — is also irrational.

There have been many such shocks since, in the development of our understanding of numbers. They have always been abstract, as was pointed out at the beginning of this section, but we are still learning the extent to which that abstraction can go. Our ability to comprehend abstraction is perhaps our greatest strength, but there is danger in distancing ourselves from the practical realm.

Rockets and planes have crashed as a result.

So called 'real' numbers include the irrational, which by definition cannot be represented exactly in any physical register, which must be composed of a finite number of digits with a finite base. While we can tolerate some error to begin with — for example, π to seven decimal places — when we combine such approximations in a complex calculation, the error can grow in a way that is hard to predict.

It is not difficult to devise a computation which ends, effectively, with a *random* number.

Which is why we should stick to (exact) integers, as far as possible, in our computer programs.

e is for Euler

Leonard Euler was arguably the most important contributor to our knowledge of mathematics. Someone once asked the legendary Pierre-Simon de Laplace what they should read to learn mathematics and he replied "Read Euler! He is the master of us all." He wrote so much (original) mathematics that even today it has not all been published. To describe him as prolific is rather an understatement.

Your calculator depends upon his work for its function.

Physics would be but a pale shadow of what it is today without it.

Among the many questions that attracted his attention, he wondered what happened when the power of a number reduced towards zero. He began with the fact that any number raised to the power zero returns unity — if you divide n^1 by n, you get one, regardless of n — and began to play.

If the power became infinitessimal, so would the excess in the result:

$$n^\alpha = 1 + \gamma \qquad \alpha, \gamma \to 0$$

Next he did something to make a contemporary mathematician cough and splutter. He raised each side to an infinite power, choosing a value to make the product with an infinitessimal finite:[10]

$$\left(n^{\alpha}\right)^{\beta} = n^{\alpha\beta} = n^{x} = (1+\gamma)^{\beta} \qquad \beta \to \infty,\ x = \alpha\beta$$

Euler made good use of the work of an antecedent — one Isaac Newton — who had generalised the Binomial Theorem to allow real numbers, leaving:

$$(1+\gamma)^{\beta} = 1 + \beta\gamma + \frac{\beta(\beta-1)}{2!}\gamma^{2} + \frac{\beta(\beta-1)(\beta-2)}{3!}\gamma^{3}\ldots$$

Since $\beta \to \infty$, we cannot reasonably distinguish it from $\beta - 1$, $\beta - 2$ *etc.*, leaving:

$$(1+\gamma)^{\beta} = 1 + \beta\gamma + \frac{\beta^{2}}{2!}\gamma^{2} + \frac{\beta^{3}}{3!}\gamma^{3}\ldots$$

allowing us to write:

$$n^{x} = 1 + \frac{\gamma}{\alpha}x + \frac{1}{2!}\left(\frac{\gamma}{\alpha}\right)^{2}x^{2} + \frac{1}{3!}\left(\frac{\gamma}{\alpha}\right)^{3}x^{3}..$$

The next leap is so simple, yet its audacity is breathtaking. Our dear Leonard could see that there must be some value of n — call it e — such that $\gamma = \alpha$:

$$e^{\alpha} \triangleq 1 + \alpha \qquad \text{definition of } e$$

No one seems to know why he chose to term it e, but our calculators can determine its value:

$$e^{x} = 1 + x + \frac{1}{2!}x^{2} + \frac{1}{3!}x^{3}\ldots$$

$$\therefore \quad e = 2 + \frac{1}{2!} + \frac{1}{3!}\ldots = \sum_{r=0}^{\infty}\frac{1}{r!}$$

There is one more way to calculate e which proves useful later. From the definition above:

$$e^{\alpha\beta} = (1+\alpha)^{\beta}$$

If we now impose the constraint $\alpha\beta = 1$, we then find:

$$e = \lim_{\beta\to\infty}\left(1 + \frac{1}{\beta}\right)^{\beta} \qquad \text{alternative definition of } e$$

which is now taken as the *official* definition.

10 There is not just one infinity. For example, there is an infinite number of integers, yet many more 'real' numbers (which include both the rational and irrational). Some infinities are bigger than others, perhaps infinitely so. Hence, it is quite an assumption that the product of an infinity and an infinitessimal is itself finite. One must choose carefully!

Differentiation: measuring the rate of change

When we draw a graph of a function $y(x)$, we're depicting how one quantity varies with another; for example, distance with time. Suppose we want to know how the rate of change varies; for example, how speed varies with time. Graphically, we could carefully measure the gradient of the line at each point of interest. But if we have some algebraic expression for y, it would be nice to derive one for the gradient also. We could then *calculate* it for any point of interest.

If we pick a point x' on the x axis then the gradient is approximated via:

$$\left.\frac{\delta y}{\delta x}\right|_{x'} = \frac{y(x'+\delta x) - y(x')}{\delta x}$$

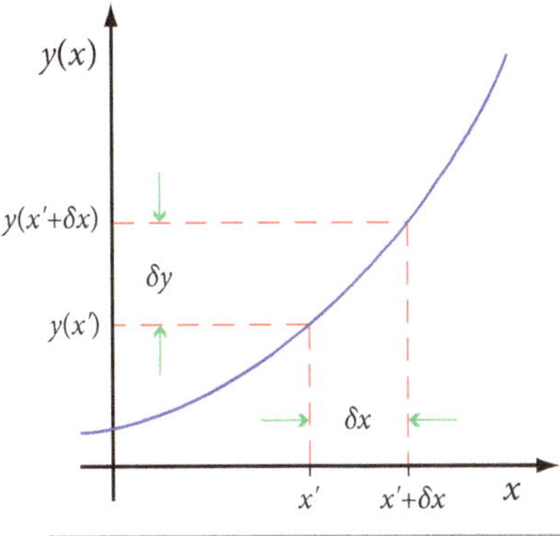

Figure 3.5 The gradient of a function.

Suppose $y(x) = x^n$. The Binomial Theorem can then help:

$$\frac{\delta y}{\delta x} = \frac{(x+\delta x)^n - x^n}{\delta x} = \frac{x^n + nx^{n-1}\delta x + n(n-1)x^{n-2}\delta x^2 + \cdots - x^n}{\delta x} \qquad n \neq 0$$

$$= nx^{n-1} + n(n-1)x^{n-2}\delta x + n(n-1)(n-2)x^{n-3}\delta x^2 \cdots$$

dropping the prime, since we might have chosen any x.

Now we apply the master stroke: we simply consider the limit as $\delta x \to 0$:

$$\frac{dy}{dx} = \lim_{\delta x \to 0} \frac{\delta y}{\delta x} = nx^{n-1} \qquad n \neq 0$$

A little reflection makes it plain that we can now 'differentiate' *any* polynomial.

A similar approach allows recovery of the 'derivative' for many (but not all) functions.

Integration: measuring the rate of accumulation

Like the gradient, the *area* under a graph also often has a clear meaning. For example, if we plot speed (vertical axis) versus time (horizontal axis) then the area beneath represents the cumulative distance travelled. We might divide that area into many little columns of width δt, whose height is the average speed over that interval. To remove uncertainty over speed at time t, we can let $\delta t \to 0$:

$$D(t') = \lim_{\delta t \to 0} \sum_{t=0}^{t=t'} s(t)\delta t = \int_0^{t'} s(t)dt \qquad D(t') = \text{distance travelled up to time } t'$$

The process is called 'integration', and the result an 'integral'. '∫' is called the "integral sign" and is intended to convey summation of a different kind to that indicated by the usual Greek 'Σ' (Sigma).

A general formulation of integration might use $A(x')$ to indicate the area beneath the curve, from some initial value of x up to x', and $y(x)$ to indicate the dependent variable (varying along the y axis):

$$A(x') = \int_{-\infty}^{x'} y(x)dx$$

Here, we suppose integration to begin where $x \to -\infty$, which can yield a finite integral, as long as $y \to 0$ at the same time.

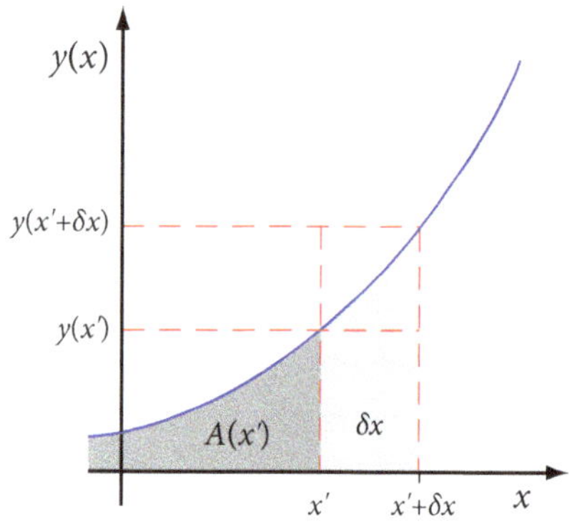

Figure 3.6 The integral of a function.

Something rather surprising emerges if we consider a small extension beyond x'. The additional area, $A(x'+\delta x) - A(x')$, must lie between that of two rectangles, of height $y(x')$ and $y(x'+\delta x)$ respectively:

$$y(x').\delta x \leq A(x' + \delta x) - A(x') \leq y(x' + \delta x).\delta x$$

Dividing throughout by δx:

$$y(x') \leq \frac{A(x' + \delta x) - A(x')}{\delta x} \leq y(x' + \delta x)$$

then taking the limit as $\delta x \to 0$, while dropping the prime:

$$\frac{dA}{dx} = \lim_{\delta x \to 0} \frac{A(x + \delta x) - A(x)}{\delta x} = y(x)$$

This is sometimes called *The Fundamental Theorem of Calculus.*

It leaves us with a method for solving the problem of integration: we must find the function whose derivative is the one we wish to integrate. We then have the 'indefinite' integral, as a function of x, which may be rendered 'definite' through evaluation at each bound, followed by subtraction.

But it also tells us something profound.

Integration and differentiation are the *inverse* of each other.

One minor problem remains when 'performing' integration.

Suppose we are seeking the indefinite integral of x^2. The function whose derivative is x^2 is:

$$A(x) = \frac{x^3}{3} \qquad \therefore \quad \frac{dA}{dx} = x^2$$

The trouble is that any constant added to the cube, contributes nothing to its derivative, so:[11]

$$A(x) = \frac{x^3}{3} + C \qquad \therefore \quad \frac{dA}{dx} = x^2$$

C is called the "constant of integration". To find it, some additional information is needed.

Differentiating a logarithm

The trick for finding the integral of any power, and thus of any polynomial, fails in one particular case: when the power concerned is –1. Thus the indefinite integral of $1/x$ remains a mystery.

Or it did until Leonard Euler, the "master of us all", arrived.

The problem he actually addressed, which gave the answer, was different. It was that of how to differentiate a logarithm.[12] While the base of the logarithm matters little — there is a simple formula for changing base — it was only 'natural' that he should choose base e. After all, that was *his* number.

Applying the ideas we gave earlier, as the "first principles" of differentiation:

$$\frac{\delta y}{\delta x} = \frac{\ln(x+\delta x) - \ln(x)}{\delta x} = \frac{1}{\delta x}\ln\left(\frac{x+\delta x}{x}\right) = \ln\left(1+\frac{\delta x}{x}\right)^{\frac{1}{\delta x}}$$

$$\therefore \quad \frac{\delta y}{\delta x} = \ln\left(1+\frac{1}{t}\right)^{\frac{t}{x}} = \frac{1}{x}\ln\left(1+\frac{1}{t}\right)^{t} \qquad t = \frac{x}{\delta x}$$

$$\therefore \quad \frac{dy}{dx} = \lim_{t\to\infty}\frac{1}{x}\ln\left(1+\frac{1}{t}\right)^{t} = \frac{1}{x}\ln\lim_{t\to\infty}\left(1+\frac{1}{t}\right)^{t} \qquad \text{as } \delta x \to 0,\ t\to\infty$$

$$\therefore \quad \underline{\frac{dy}{dx} = \frac{1}{x}}$$

As a result, we now also know the integral of $1/x$:

$$\int \frac{1}{x}dx = \ln(x) + C$$

Just what we need to derive the Rocket Equation!

---

11 The derivative of a sum is the sum of the derivatives of the terms, and the derivative (gradient) of any constant is zero. It can also be fairly easily shown that the integral of a sum is also the sum of the integrals.

12 A 'logarithm' (or 'log') is just the power to which you must raise some 'base' to obtain a given number. Thus the logarithm of the base is always 1. Applying the laws governing the combination of powers yields the laws of logarithms. Hence, for example, $\log(a.b) = \log(a) + \log(b)$, and thus $\log(a^n) = n.\log(a)$. If the base is e, the log is called 'natural' and written 'ln'.

Chapter 4

Computers

4.1 Introduction

What is a computer?

Mechanical computation

The term 'computer' used to apply to a human. Computation was a profession. As a result it was slow and expensive, and thus limited in application. Only wealthy merchants and the occasional government could afford to indulge in it. Engineers, scientists and navigators appealed in vain for help from mathematicians, who were usually interested in higher things than mere calculation.

It may seem hard to believe that harm can come from merely getting a sum wrong but indeed it can, and still does to this day. As trade boomed in the early 19th Century, more and more life was lost on the high seas as a result. The cause could often be traced to a mistake in a logarithmic or trigonometric table used when determining position and heading from astronomical observation. Each entry had to be laboriously calculated by hand, and it was all too easy for an error to go unnoticed.

Charles Babbage, Lucasian Professor of Mathematics at Cambridge University, believed that the answer was to *automate* their production, thus removing even the possibility of human error. As one of the foremost mathematicians of his day, he could see how this might be achieved.

It might seem surprising to us that anyone might even contemplate automation so long ago, but the Industrial Revolution was founded upon it. Human labour was being replaced by machine everywhere back then, most especially in the production of textiles. The Jacquard Loom (1839) was even *programmable*, to replicate the design of any pattern, which undoubtedly gave Babbage inspiration.

He designed a machine called the "Difference Engine" which would perform a given computation when the handle was cranked. Although it was possible then to machine its parts to the required

Figure 4.1 Charles Babbage.

precision, he never saw it completed. Because it was limited to a single calculation at a time, it cannot be called a 'computer'. It could assist a human computer but could not do the job alone.

A working Difference Engine was built by the London Science Museum and its neighbour, the Department for Mechanical Engineering of Imperial College, between 1989 and 1991.

Mechanical calculators had been designed and built before. We now know that the Ancient Greeks made machines which would predict the movement of planets centuries before the birth of Christ.[1] The rotation of a gear formed the *analogue* of the quantity of interest. Babbage's engine was *digital*. A distinct orientation of each gear marked the value of each decimal digit. But was it right to use decimal?

We are left with two questions. First, is a digital approach better than analogue and, if so, why? Second, what base should we employ in building a device to register a value?

We must address each of these problems before we can proceed.

Charles Babbage arguably has more claim to be the father of the artificial computer than anyone else. Dissatisfied with the Difference Engine, he designed his "Analytical Engine" to be capable of completing any sequence of *conditional* computations, according to a designated program. It was a monster, measuring some fifteen metres long and five high, and powered by a steam engine. Had it ever been built, it would have been the world's first programmable digital computer.

It is truly astonishing how many great inventions were made in Britain. Only slightly less astonishing is the British failure to exploit them commercially. The two remaining tricks necessary for the remaking of our modern world also originated here: the ability of the machine to store its program within its own memory, and the use of electronics, for speed and efficiency. The world's first all-electronic, stored-program computer was the "Manchester Baby", born in Manchester in 1948.[2]

Today, there is no British company remaining which manufactures computers.

But there is a very successful one that designs them (ARM).

There are more 'firsts' for British computing. Babbage knew a young lady who repeatedly astonished him with her brilliance. Of course, being a woman at the time rather put paid to any hope Ada King, Countess of Lovelace, might have had of an academic career, or one of any other kind. (You wonder how much has been lost through such cruel misogyny.) Nonetheless, hers is one of the most wonderful accomplishments in the history of technology: she published the world's first computer program.[3]

But even that palls when compared to her next contribution.

1 Sponge divers, off the island of Antikythera, recovered one in 1901 from a Roman shipwreck.

2 Built by ex-radar physicists who reportedly requisitioned all necessary parts from their former employer.

3 It must be added that Babbage *wrote* the first computer programs, but did not publish them.

Why is it so important to automate computation?

Figure 4.2 Ada Lovelace.

Ada had enjoyed an excellent technical education, with tutors including the renowned London mathematician Augustus De Morgan. She understood the relationship between science and mathematics. This allowed her to notice that *everything* can be reduced ultimately to numbers.

At first, this seems a rather paltry observation, but that is only because we are quite used to digital cameras, telephones, televisions *etc.*, and 'digits' are just the elements which make up numbers. In 1842, it was nothing less than momentous and prescient. She could see the potential for everything that we now take for granted, and that defines our world and how we function within it.

> ... it might act upon other things besides *number*, were objects found whose mutual fundamental relations could be expressed by those of the abstract science of operations, and which should be also susceptible of adaptations to the action of the operating notation and mechanism of the engine.[4]
>
> *Ada Lovelace*

Her note also captures the idea of *symbolic* computation — that the machine might manipulate 'objects' that represent something other, yet more fundamental, than numbers. This is the very essence of the mathematical model proposed for computation by Alan Turing, a century later.

We are now free to regard the computer as the means by which any 'system' might be implemented, merely by determining the program which correctly captures those 'relations'.

A 'system' is anything that has both 'input' and 'output' and defines the relationship between (Figure 4.3). Both input and output may change over time, and thus each forms a stream of symbols.

To build a system of any complexity, one must reduce both the task and the system. This distinction is critical, and has long been understood in the world of mechanical, and civil, engineering. But the engineering of computer programs — or 'software' — is a much younger art, and far less mature.

We reduce a system to a network of components, each of which has a well-defined interface to its neighbours. With reference to language, a 19th Century philosopher called Gottlob Frege introduced the idea of 'compositionality', under which any component can be replaced with another sharing the same specification

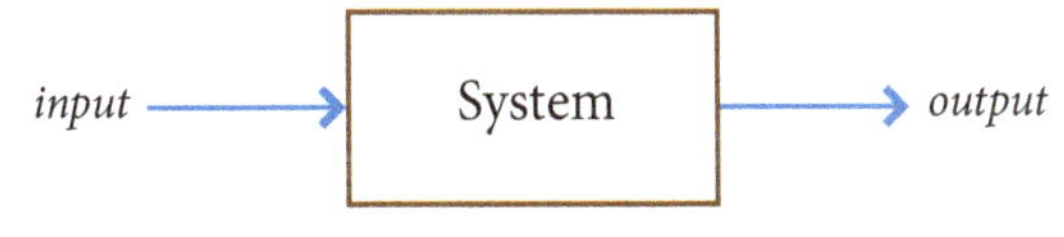

Figure 4.3 A 'system' as anything with input and output.

4 From *Note A* in the English translation of *Sketch of The Analytical Engine Invented by Charles Babbage with notes upon the Memoir by the Translator*, Scientific Memoirs, **3**, 1842.

without altering the behaviour of the system as a whole. This is a principle which underpins, if not defines, any mature engineering discipline. It requires an adequate language in which to describe an adequate specification. That language ideally should afford composition of component specifications, ultimately allowing a full and precise description of the system.

The engineering of computer hardware is indeed compositional.

Sadly, that of software is yet rarely so.

Reduction of the *task* is to meet different concerns. What might be termed "project modularity" (as opposed to "system modularity") serves the needs of separated development and reuse (as opposed to construction and maintenance). This is not the place to explore how this is achieved, with regard to either hardware or software. Suffice it to say that this too is an area of difficulty with software.

What the computer brought to system development, in a very wide and still expanding area of application, is the ability to render the hardware common, leaving software to define function. The introduction of the microprocessor, in particular, allowed mass production to dramatically reduce system cost, while permitting the function of a product to expand and vary. The culmination of this has been called *ubiquitous* computing, made manifest in the contemporary tablet or phone, and the fact that even a toaster is probably controlled by its own *embedded* computer. A typical car these days may have fifty computers embedded. We have yet to explore what happens when we interconnect them, or when we allow the appliances themselves to connect.

There are doubtless risks, as well as opportunities.

We have already noted that the engineering of software is a young and immature discipline, despite what many academics and professional programmers might protest. Though the first published program dates back a hundred and seventy years, it took another century before a computer even existed.

To this day, there remains considerable resistance among student and hardened professional alike to complete and proper engineering discipline. Despite software taking its place alongside mechanical and electronic systems, in the development of, say, a car, it is still common to hear remarks about its need for special dispensation of one kind or another. But when you run a business and sell things, and are held accountable in every way, you *must* have properly conducted and *documented* specification, design and test (among other things).

However, software *is* more complex than any mechanical or electronic system. As anyone who has tried programming will testify, it is hard enough to get it right, still harder to *prove* that you have. If the lives of companies and even people depend upon correctness, proof is essential. The necessary knowledge and skills exist, but are currently rare and thus expensive to recruit. Without effective tools for design and development, our ability to prove correctness will not scale with the size of project.

Mechanical engineering once suffered the same problem. For decades, rail travel involved a significant threat to life and limb. It was even more dangerous than road transport is today. A small group of unsung heroes eventually found ways to make it safe. The safety of air travel is only possible because of their work. Arguably, the greatest risk in using any mode of transport now is due to its growing dependence upon software. Hopefully, we will soon learn how to render software safe.

One approach is to generate it automatically, directly from a specification. But this still leaves two problems: the generator had better be correct, as well as universal, and so had our specification. It may prove no easier to get our specification right than it was a dedicated program.

Wonderful, ubiquitous and cheap computers may be, but safety remains an ambition.

Ada correctly emphasised that, despite appearances, the machine would never embody any *understanding* of its function, but would merely execute its instructions mechanically, as an automaton. Today, there are those who would disagree, and who believe in "artificial intelligence". They maintain that if it performs an 'intelligent' function then it must be credited with intelligence. This is not the place to expand on this dispute, except to say that it obviously reduces to the definition you choose for the term 'intelligent' (which is used to describe many disparate things) and your point of view.

There should be no doubt though that machines will continue to expand their powers. They are progressively displacing labour, even skilled labour. While this process, as in the previous industrial revolution, is causing mass unemployment, the longer term benefits are profound. Every successful civilization depends upon slavery, of one form or another. Machines are morally acceptable slaves. They also take no holiday or tea break, and get sick rarely. They neither tire nor rebel.

There is now no process we cannot automate.

The days of human labour are drawing to a close.

Prerequisites

There is a myth that, in order to understand how a computer works, one must first understand electronics. We have already spoken of the purely *mechanical* computer, which Babbage envisaged, so this is clearly untrue. Indeed, in the future, we might employ yet another medium, such as light.

It turns out that there is a definite divide between how a computer is constructed, and how the elements from which it is built are made. Those elements are merely switches. Here, we take them for granted. We leave their fabrication for another time or place, or for someone else to worry about.

Before we start talking about how switches can be interconnected to form a computer, we do need to know a little bit about information, memory, communication and logic. We also have those two issues to resolve: why digital, rather than analogue, and what base ('radix') should we use?

Information and memory

What is information?

It would seem a good idea to start by saying precisely what we mean by 'information' :

Information is that which resolves uncertainty.

In particular, information allows us to decide between a number of alternatives. (Note that communication is always implied.) We are left with three problems.

The first problem is semantic. The meaning of any message clearly depends upon the set of alternatives over which we are uncertain, which must be identical for both sender and receiver. Otherwise, the consequences of any transmission will be unpredictable, negating any purpose in its undertaking. A 'protocol' — a set of conditions agreed prior to communication — is always necessary.

The second problem is philosophical. Information is a little more than merely establishing an alternative from a given set. When we examine a signal, we are capable of deeper interpretation. For example, a detonator may be armed, or the order received to "open fire". Meaning requires sentience.

Only humans, and other sentient creatures, experience meaningful uncertainty.

An engineer is not (during office hours) concerned with philosophy, and will usually prefer to leave this to others. As a result, they tend to refer to messages as 'data', and not as information. They may provide you with an efficient channel over which to send much data. If your messages contain no meaningful information whatsoever, that's just fine, as long as you pay your bill.

An ancient illustration of the difference between data and information is that of the politician, who must speak often but say as little as possible, lest they be held to account. It is possible to show they must lie only *half* the time, otherwise their audience will either ignore everything they say, or acquire information via inversion. For something simpler, go to an electrical store and buy a "four-gang" light switch. The configuration of the switches could be described using four binary digits (bits). Once wired into your house to control four lights, that data might be said to 'contain' information.

Measuring information content

The third problem is how to measure information, and in what units. Clearly, the larger the set of possible messages the greater the quantity of information conveyed. Hence, any measure should increase with set size monotonically. The credit for proposing a *logarithmic* variation is due to R. V. L. Hartley of the Bell telephone Laboratory ("Bell Labs") in the 1920s.

To see why this makes sense, consider a channel consisting of a set of three decimal thumb-wheels like those commonly used to lock a briefcase. If a fourth is added, there will be ten times the number of alternative 'messages'. The set size varies exponentially with the number of wheels. Yet, in such circumstances, it is usually desirable to speak of the quantity of information conveyed as varying linearly. If we define it to vary as the logarithm to base ten ($\log_{10}$) of the set size then adding a single wheel simply increments the information content of each message by one unit. The decision to employ a logarithm is arbitrary and is taken just to make things intuitive and convenient.

The choice of base for the logarithm is also arbitrary but is perhaps easier to understand. Suppose there are just two alternative messages. Using decimal, a single transmission would transmit $\log_{10}2 = 0.301$ decimal units. Somehow, speaking of less than one unit seems unnatural. Using binary (base two), we would speak of one unit. There can never be fewer options than two to choose between in receiving any message. As a result, we need never speak of less than one unit of information if we use binary. For this reason, the binary unit — 'bit' — of information quickly became universal.

We shall see later that binary is not just natural but both efficient and convenient as well.

It should be emphasized that the term 'bit' denotes two quite different things. As we have seen, one is a unit of information such that, on receipt of a message, two alternatives may be distinguished — for example, the red and green flags used at a railway station. 'Bit' is also a contraction of "binary digit", as employed when writing down a number using binary notation. Of course, a binary digit *can* convey a binary unit of information. It might carry less, though never more.

State and value

State refers to the configuration of some physical system; for example, a planet in orbit around a star. Any system capable of more than one distinguishable state can act as a memory device. Given the necessary superhuman powers, you could use a planetary orbit to remind yourself of something.

A more reasonable example would be a triangular bar resting on a table (Figure 4.4). This has three stable states (if we discount standing it awkwardly on its ends). To turn this into a satisfactory memory device, all we have to do is *label* each stable state. Perhaps we might use the system to remember who is manning a help desk, by writing one name on each surface. In summary, memory is the labelling of at least two distinguishable physical states.

Figure 4.4 Memory.

Only the label is important to the user of the device. The particular state employed in each case is unimportant. Sometimes, the user is able to write a 'value' (label), as well as read one. A computer program makes use of many such 'registers', which are perceived as locations within a linearly consolidated memory, each one accessed via an address, like a house on a street.

It will prove essential to keep separate the concepts of state and value. States may be simply enumerated. Each value constitutes an interpretation of a particular state, and possesses meaning.

In this book, we are more concerned with implementation than usage. Hence, it's worth saying more about both state and the consequences of changing it.

There are three distinct kinds of stable state :

— static no change with time

— periodic cyclic change, *e.g.* a planetary orbit

— chaotic acyclic change (unique trajectory but without periodicity).

A static state can be thought of as periodic, but with period zero. It is thus possible to argue that there are only two kinds. Though inappropriately named, the discovery of 'chaos' marks one of the greatest breakthroughs in physics of the 20th century. (A better name would be 'aperiodicity'.)

To date, all digital memory has been built using static states. However, it is now appreciated that nature employs periodic states in the neuronal networks that make up a brain. It is quite possible that, as such systems and their advantages become better understood, that artifice might follow nature.

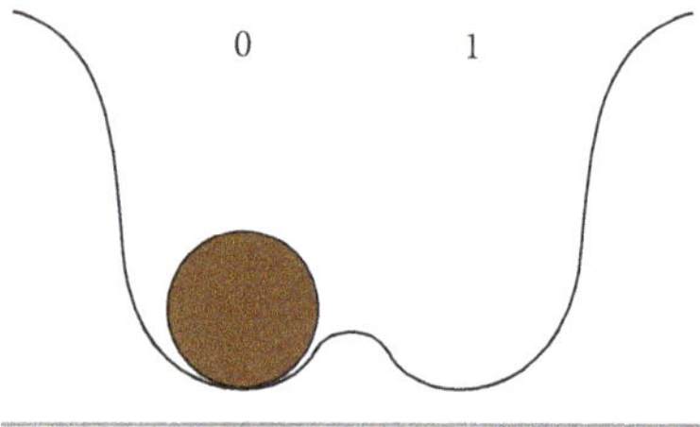

Figure 4.5 One of two stable states.

A system in a stable state need consume no energy. The reason why a computer consumes energy is that memory is continually being updated; *i.e.* states are being changed. It is impossible to construct a device which does not consume energy when changing its state. Consider a ball in a well with a ridge in the middle (Figure 4.5). Clearly, there are two stable states, which we can label however we choose. The higher the ball, the greater is its potential energy (or just 'potential'). Force must be applied to move the ball up and over the ridge in order to change state. Hence, work must be done and energy expended. This is recovered when the ball rolls down the other side. For the ball to come to rest, something must absorb that energy. The only possibilities are the ball and the well, one or both of which grow warmer. Physics tells us that, not only must a working memory consume energy, but also that it will end up dissipated as heat. As heat accumulates, temperature rises, and some balance will be established. This is achieved by a combination of radiation and conduction.

The long and the short of all this is that your lovely new laptop computer must have a battery, which must be regularly recharged somehow, and some means of cooling. The greater the rate at which state is changed, the more likely it is this will be a fan or heavy radiator of some kind. Physics matters. It has tangible consequences. A computer can overheat. People have been burned by their laptops.

There is, at least, one limit to how low the power consumption of a useful device can be. The barrier between states must be high enough to prevent change due to background 'noise' (see p.89). The higher the barrier, the greater is the energy required to change state. We have yet to approach that limit.

It is important to understand just how dependent the apparent progress in information technology is on that in physics and its application. There is much still to come.

The revolution that took place in physics in the early 20th century has still to play out.

Another way to contrive a 'bistable' device is to take a glass and draw a line halfway up. Two states can then be distinguished; either the glass is full or empty, which can be decided by whether the content rises above or falls below the line. We might label these states by drawing a '1' above it, and a '0' below it. To store the value '1', we fill the glass. To store a '0', we empty it. In order to read the correct value, the decision must not be in doubt. Uncertainty must be resolved.

Now suppose the container leaks. All may not be lost. First we must determine just how leaky it is — how long does it take to lose half? Provided the leakage is predictable, we're still in business. We must periodically *refresh* the state. Before the leakage becomes a threat, we determine the state and then reset it fully. If the glass is empty, we need do nothing. If the water level is above the line, we must refill the glass. Some "sense amplifier" is necessary for any such *dynamic* memory device (Figure 4.6).

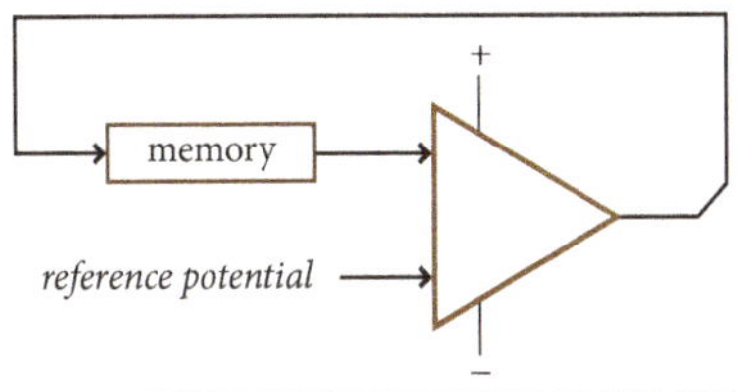

Figure 4.6 Refreshing a dynamic memory.

It may seem a bizarre way to make a memory device but it illustrates a method commonly employed in digital systems and computers. Glasses of water would be a little awkward. An electronic 'capacitor' is used in place of the glass, and charge in place of water. An electric current is just a flow of charge, which can be accumulated in a capacitor.

Capacitors are not just easy to make on a semiconductor substrate, they are hard to avoid. A very large number can be accommodated within a very small area. The problem is that they do indeed leak. Not only can charge leak out, it can leak in as well, depending on what is happening in the cell next door. As luck would have it, the "refresh period" required is around a millisecond, which is a very long time compared to the typical interval between memory 'accesses' in a computer.

A so-called 'dynamic' memory need only rarely be inaccessible.

Register and base

To record the choice between M alternatives, and thus store $\log_2 M$ bits (binary units) of information, we could use either :

- one device with M (distinguishable and stable) states, or
- n devices, each with k states.

Either system is known as a "register of modulus M" and implements a distinct unit of memory whose state may be represented by a word of width (or, if you prefer, length) $n = \log_k M$. A 'word' is a sequence of digits with base k. For example, many computers employ binary memory with word width thirty-two.[5] We thus call them "32-bit machines". It is sensible to make all registers the same

5 64-bit machines have also become common.

width. Otherwise, simply copying a value from one register to another can involve additional operations; for example, to correctly represent a negative value.

Nature performs rather a lot of computation, and thus requires a number of different memory systems. For example, evolution records the way in which a 'phenotype' is grown via a 'genotype' — a particular genetic pattern. Genetic memory is linear and ultimately composed of paired chemical 'nucleotides'. A 'chromosome' is a single very long molecule, rendered compact by coiling up into a double helix. To be read, it must be unwound to form two long chains in opposition.

There are just four kinds of nucleotide — termed A, G, C, and T, after their chemical names — and thus four possible pairings. Nature is complicated. It may have special reasons for employing a quaternary (base-4) genetic memory instead of the binary now universally used in computers. We do not know. It may simply be down to utility — it is easy to accomplish via organic chemistry.

We return to the problem of constructing digital memory and ask "why binary?" Would some other base be better? To answer, we must first decide what we might care to optimise. Clearly, there are two concerns: speed and cost. We would like to be able to both read and write a register as fast as possible, and we would like the cost of each one to be a minimum. When the first computers were built, the cost was astronomic. Keeping it under some sort of control was essential. Today we care much more about speed, but we also desire a lot more registers — hundreds of millions of them. Hence, the cost of each one is still important. As a result, we shall attempt to determine the value of k that minimizes cost, and worry about speed later.

Suppose you are washed up on a desert island and, unlikely as it may seem, need to make a memory device, *i.e.* a register. Perhaps you would like to mark time within a calendar month. Let us say you can count off the days two at a time. Hence, you need to distinguish between sixteen different periods. With a bit of ingenuity you can make a primitive adze, with which you can gouge indentations in a block of driftwood. One solution is to gouge sixteen holes in a single block and mark one with a pebble (Figure 4.7a). Thus you have made a register of modulus sixteen comprising a single device. By far the greatest cost has been our time. (Pebbles and blocks of wood may be assumed plentiful.) If we measure cost according to the time required to carve a single indent, the cost has been sixteen units.

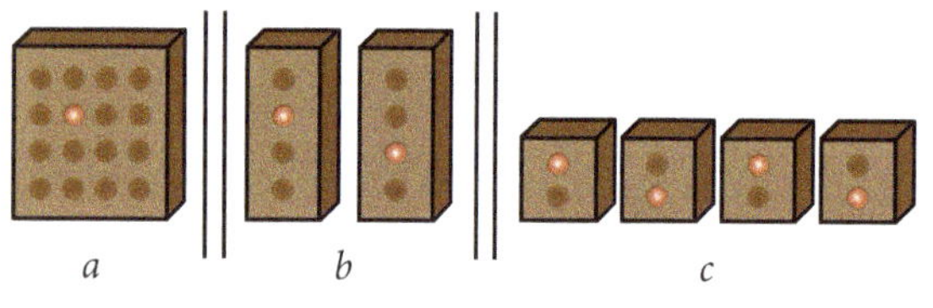

Figure 4.7 Three registers with modulus sixteen.

The alternative is to combine multiple devices. There are two options that yield exactly the same number of configurations. We could carve four indents in each of two blocks (Figure 4.7b) or two indents in each of four (Figure 4.7c). If we don't count the extra blocks and pebbles, the cost is the same in each case, at eight units. On the other hand, if we do allow some small account to be taken of blocks and pebbles, say in the time needed to collect them, then the last solution, using binary devices, is slightly more costly than using four indents per device. Quaternary wins by a small margin.

Note that the state of any register can be written as a sequence of digits of corresponding base. A hexadecimal (base-16) digit can be written using ordinary decimal characters for zero to nine and the first six letters of the alphabet for ten to fifteen. Thus we can label each indent in the single block of our first solution. For example, the tenth state/indent would be labelled 'A'. Where multiple devices make up the register, we use one digit per device. Hence, two quaternary digits ('0'…'3') describe the state of the second solution, and four binary digits (bits) define that of the third.[6]

We will now take a deep breath and give a formal analysis of the cost of making a register of modulus M with regard to radix (base) k. Exactly the same analysis may be applied to the cost of making a communication channel. You need either $\log_k M$ memory devices or $\log_k M$ cores in your cable.

The only assumptions made are that the cost is linearly proportional to the number of devices making up the register, and that the cost of each device is similarly proportional to the number of states of which it is capable. Hence, the total cost varies as the product of the two, and we can write:

$$cost \propto k \times \log_k M$$

It's convenient to change to a standard base for the logarithm. Here, we choose base-2:

$$cost \propto k \times \frac{\log_2 M}{\log_2 k}$$

The term '$\log_2 M$' is a constant, independent of k. Hence, it can be absorbed into a new constant of proportionality, leaving exposed the variation in cost of any register with radix:

$$cost \propto \frac{k}{\log_2 k} \qquad (1)$$

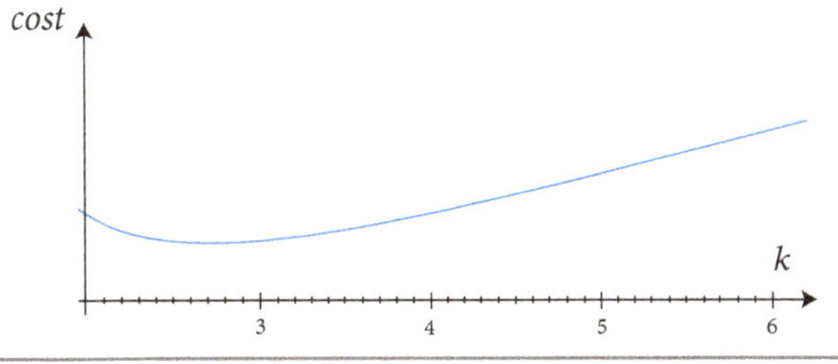

Figure 4.8 Variation of register cost with radix (base).

Figure 4.8 shows this variation graphically. The form taken is the same regardless of register modulus. (Varying M merely moves the curve up or down.)

If the reader is wide awake, they will have spotted something curious. The minimum cost is *not* achieved with binary (base 2). Ternary (base 3) is actually cheaper, but only slightly.[7]

Here, we enter the land of giants — those who made the breakthroughs that led to our era of digital systems. Norbert Wiener, of the Massachusetts Instutute of Technology (MIT), published a

6 Take care when using the term 'bit'. Recall that it has two distinct meanings: a binary digit, or a binary unit of information, sufficient to resolve between two alternatives.

7 The lowest point of the cost curve is actually at $k = e$ (Euler's number), but we need a whole number.

little book in 1948 entitled *Cybernetics* that was principally about automatic control systems.[8] However, it also addressed fundamental issues in digital communication and computation, including the best way to construct a register. Wiener made essentially the same assumptions we made above but with one essential difference (Chapter 5). He assumed that the cost of each device, that recorded a single digit, would vary not as k, but as $k - 1$. In other words, he argued that one state could be gained for free. We shall call this the "Wiener assumption".

For his register, he proposed the use of a "uniform scale", with some sort of marker moving along it. You record a digital measure by leaving the mark in one of a number of distinguishable regions. (Imagine them each painted a distinct colour.) His argument was extremely simple. To record one choice from k possibilities, you need only paint $k - 1$ regions. The final possibility is distinguished when the mark lies outside any of these. The point is that we have only needed to meet the cost of $k - 1$, not k, regions, although we still need the same number of devices: $\log_k M$.

Hence we can write:

$$cost \propto \frac{k-1}{\log_2 k} \qquad (2)$$

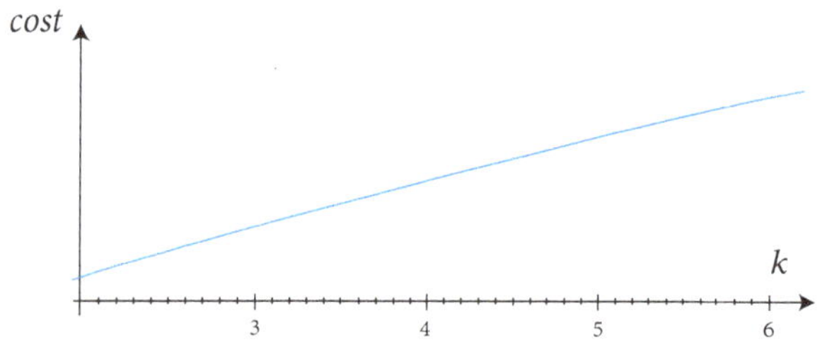

Figure 4.9 Register cost, accounting for Wiener assumption.

We can see from Figure 4.9 what a difference this makes. If the Wiener assumption holds then the cheapest way to build registers is by using binary devices. But does it hold?

It depends upon the device you use.

Consider the block and ball memory discussed earlier. Norbert Wiener is right! It is indeed possible to get one state for free. As Figure 4.10 depicts, we need only carve, for example, three indents to achieve a base-4 device. The first three states are contrived by placing a pebble in one or other indent. The fourth state is achieved by simply not placing the pebble in *any* of them. If the cost can be truly measured, or at least reasonably approximated, by counting indents then binary now has a clear advantage.

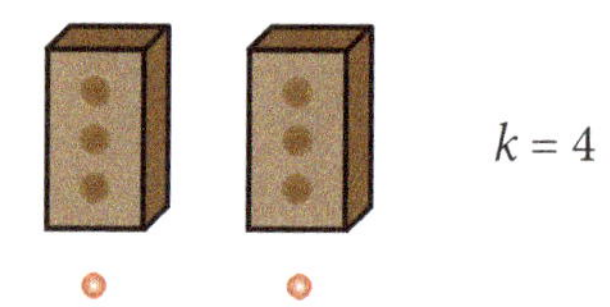

Figure 4.10 A cheaper register.

We worked too hard before.

This is the true dominant reason why we build exclusively in binary now. Other explanations are sometimes wrongly given. For example, it has been said that binary is used because it relates directly to a logical calculus, *e.g.* Boolean algebra (see later). However, it is possible to devise a *three*-valued calculus that would allow the implementation of arithmetic operations. It has also been

8 The title is derived from the ancient Greek 'kybernetes', meaning helmsman.

said that binary was chosen because the unit of information is also binary, *i.e.* to equate the latter with a binary digit. But the adoption of that unit is entirely arbitrary (though convenient).

One should be clear: the reason digital systems employ binary representation is because *it minimizes their cost.* It seems obvious now, but that is often the mark of genius. Once someone with the intellectual might of Norbert Wiener clarifies an issue, it often seems obvious afterwards. Only a fool would dismiss the contribution, or diminish its importance. Yet many decisions concerning our technology are now taken for granted, and are thus frequently misunderstood.

During the Cold War, Soviet science and technology largely kept pace with that of the West, and sometimes exceeded it, despite the fact that the two sides enjoyed little communication. It should not be too surprising then that one side proposed (in the USA), and the other actually built (in Russia), a machine based on the same principle, but each in complete ignorance of the other.[9] It might be more surprising to learn that the principle concerned was the storage of information using base three.

Not all memory devices obey the Wiener assumption. An example, that we shall meet later, is the use of two switches to record a single bit. Arguably, the decisive reason why ternary machines never caught on was that they would save only a small proportion of the cost of a binary alternative, and then only if devices were used which denied the Wiener assumption.

A bigger saving could be made by using binary devices in accord with it.

Before we leave base-3 behind, *balanced* ternary notation is worthy of note. Instead of enumerating cardinal (unsigned) integers using the symbols '0', '1', and '2', we instead allow a *vinculum*, or 'overbar', to denote arithmetic negation of the unit in any position. This allows us to write down signed integers using only the digits '0' and '1'. For example:

$$1\bar{1}0\bar{1} = 17_{10}$$

Each digit represents a power of three but with the chance of making a *negative* contribution.

As well as forming an extremely compact notation, it affords a convenient way of representing signed integers within a ternary register, since no separate record of sign is needed. It is also efficient in communication, since any form of current or potential can be reversed.[10] The idea can be traced back to one of many great French mathematicians, Augustus Cauchy.

You may be surprised to learn that it is possible to write numbers with just one symbol (and no decoration) — a *unary* notation. Your ancestors would not be. The vast majority of trade throughout

9 The US proposal was for the historically important Whirlwind project at MIT, and the Russian design was for the Setun series of machines built by Nikolai Brusentsov *et alumni* at Moscow State University. Unfortunately, the Russian machines failed to save any money because they actually used binary (magnetic core) devices, perhaps because they were easier to make or just easier to buy. Setun also reportedly made use of balanced ternary representation.

10 Each digit is transmitted via a channel with three states: flow/potential forward, back, or below some threshold.

history has been conducted using unary. A typical device employed was the "tally stick". You simply cut notches in a stick to record how many bushels of wheat, pots of wine, *etc.* were owned or owed.

For small numbers, the tally stick works well. Local trade demanded no more. Emperors, and their rich merchants did need more, however. It takes time for your slaves to carve the record of a thousand casks of olive oil, and even their time is worth money.

Suppose you find yourself in a prison cell and wish to count the days by marking the wall. The eye can easily reckon the number of closely scored marks up to a point not far beyond five. So you group them in fives in some way, perhaps like this :

~~1111~~ ~~1111~~ 111

As the months go by, it becomes hard to see how many groups of five there are. So you hit upon the idea of marking groups of five groups of five :

~~1111~~ ~~1111~~ ~~1111~~ ~~1111~~ ~~1111~~ ~~1111~~ ~~1111~~ ~~1111~~ ~~1111~~ ~~1111~~ ~~1111~~ ~~1111~~ 111

You become aware of the principal at work, and that one could generalize the idea to *any* power of five. It took thousands of years before people (somewhere in the East) came up with the final two ideas. First, by designing some extra symbols — in this case, four — and by attaching a meaning to their *position*, you can record the same number much more easily. The above would be written :

2 2 3

The symbols we now use came to us via an Arabic culture, which is why we write numbers from right to left, as in Arabic script. They may have originally been devised according to the number of included angles. (Try drawing them with the minimum set of straight lines.)

One problem remains : what happens if there are *no* groups of a certain power? To simply omit that column invites ambiguity and misinterpretation, perhaps with disastrous consequences.

Someone had to invent *zero*. It is critical for any positional notation, regardless of base (radix) employed. A simple space is not clear enough, when a trade dispute, or even a war, might result.

An intriguing explanation for the symbol chosen for zero relates to the "counting board" — an alternative to the tally stick. A wooden tray would be filled with sand, and pebbles added in a row. To increment the count, you add a pebble. To decrement, you remove one. When the last item was sold, and the last pebble removed, a roundish depression would remain where it had lain.

The invention, or perhaps discovery, of zero remains one of the greatest in history. Without it, there would be no positional number notation, no register, no digital system, and no computer. We would not have gone to the Moon, and would not have built the internet.

Everything arose from nothing.

Representing state and number

Now we've established that base two (binary) offers the cheapest way to make registers, it's a good idea to say a little about how you use them. First, we should note that any- and everything can be recorded as a number. For example, text is just a sequence of characters, which are encoded as distinct numbers. Even the instructions understood by a computer may be considered to be numbers.

The character, instruction or whatever, stored in a register, may be called the 'value', and the natural number given by the sequence of binary digits (bits) within it, the 'state'.[11] It is very important to keep 'state' and 'value' separate. The mapping between them is called a (simple) "data type". For example, if the values are natural numbers then the mapping may be direct.

It is often useful to refer to the state of a register, but binary is often (very) inconvenient, especially when the register width is large. Engineers like to keep things simple for a reason better than convenience: if they make a mistake, it can be very expensive, so they try to keep the chances low. Writing long binary numbers invites error. Decimal comes with a different problem: with the same number of digits, some numbers will represent a valid state, others will not. Suppose we have a four-bit register; we will need two decimal digits to cover the sixteen states, but then 17 will not refer to one.

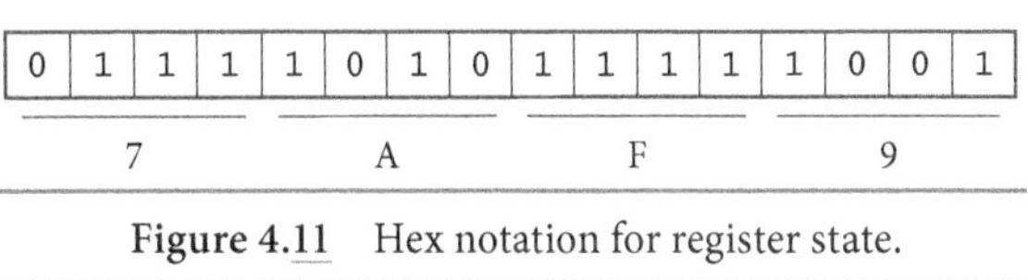

Figure 4.11 Hex notation for register state.

To, again, keep things simple, engineers design computers so that every register has the same width. As that width has grown, another convention has emerged: width remains a multiple of four bits.[12] This has prompted a solution to the efficient description of state: hexadecimal notation, or 'hex'. Each hex digit maps to one four-bit state, and *vice versa*, with no redundancy. A contemporary sixty-four-bit register thus requires just sixteen digits to describe any state.[13]

Figure 4.11 depicts a state of a sixteen-bit register and its notation.

Because numerical computation is so important, it's worth saying more about numerical data types. For example, the "dynamic range" — the set of values possible, as the system runs — of a natural number (henceforward, just a 'natural') may not mirror the set of register states. There may be an offset, and/or values may count large units (a good way to deal with values that remain big at all times). The specification of any natural number data type must include *bounds* (both minimum and maximum).

11 'State' is a term used in computer science in at least two different senses: the state of a register, and that of some process. In the latter case, 'state' refers to the set of events which may then occur. It may also refer to current output and possibly to prior events. The process concerned may be an active 'automaton' (see later) or a running program on a computer.

12 An earlier convention for register width divisible by three is worthy of a footnote. It explains, for example, why the C programming language offers the facility to specify a state using 'octal' (base eight, where each digit describes three bits).

13 A block of eight bits became known as a 'byte', which gave rise to one of four being called a 'nybble'.

Another property of a numerical data type is its 'resolution' — the difference between values represented by neighbouring states, and thus the smallest difference that can be recorded. With integer types, resolution is always constant across the range, but is not necessarily a unit (1). We should always imagine a binary point somewhere, and not necessarily after the least-significant bit (LSB). For a dynamic range between zero and one, for example, we might position it before the most-significant bit (MSB), allowing us to count small units and leaving a resolution dependent upon register width.

In a four-bit register used to represent a fixed-point fraction, the MSB represents 2^{-1} and the LSB 2^{-4}, so that is the resolution. For example, 0110 would represent 0.375, with 0.4375 the next value up.

It is also possible to allow the binary point to 'float', by recording its location. By devoting part of the register to this — effectively recording an exponent of two —it is possible to record values with a very large dynamic range indeed, according to the size of field allocated in this way. It is a profound, but lamentably common, mistake to associate a "floating-point" data type with real numbers.

Between any two real values, there are infinitely more, so it is unlikely in the extreme that any state will represent one in particular. On almost every occasion, there will be a difference between the chosen value and the one actually represented. Unfortunately, such error grows when values are combined arithmetically in a manner that is very difficult to predict. After a great many computations, it is quite possible for the result to be effectively random.

There was a time when computers were endowed with registers so narrow that the limited dynamic range of integer data types was a serious inconvenience. Sixteen-bit machines could offer only 65,536 values for any variable. Floating the point offered a way of circumventing this difficulty, while introducing a new problem that was often not properly understood — resolution would vary across the dynamic range, as the exponent varied and the binary point floated.

A thirty-two bit register offers over *four billion* values. Sixty-four-bit computers are common.

There is now rarely any circumstance where floating-point computation is justifiable.

Willing as we may be to make do with integers, and forego the full mathematical power of real numbers, *signed* integers are rather useful. So how do we map the states of a register (effectively, a bounded subset of natural numbers) to account for sign?

From here onward, purely for convenience, we shall refer to unsigned integers as 'natural' numbers and their signed brethren as simply 'integers'. (When we say 'integer', we will always mean signed.)

We need to think carefully about exactly what we need to achieve. Here's a summary:

- a unique state to represent zero
- range balanced between negative and positive values
- subtraction effected via natural number addition.

Having significantly more negative than positive values, or vice versa, clearly makes no sense. Performing signed arithmetic using the same hardware as unsigned is simpler and more efficient.

We'll dispense with any discussion of failed alternatives and simply describe the one method which achieves all our aims, and which is now employed universally (well, on Planet Earth, anyway).

The "two's complement" of a state α of a register with modulus M is defined as:

$$\alpha^{(2)} = M - \alpha$$

It earns its name because the effect, when using binary notation, is the same as subtracting each bit in succession, from LSB to MSB, from two, borrowing 1 each time from the next column up. Hence, each bit that results is the "two's complement" of the original.

We use the two's complement of the magnitude to represent the corresponding negative value.

For each α, there is just one $\alpha^{(2)}$, so we know the range is balanced. If we use $\alpha = 0$ to represent the value zero, then its complement is the very same state, and no duplicate emerges. If we add natural numbers and merely re-interpret them and their sum via two's complement notation, we find we obtain the correct answers, with just one concession: there is a new bound between positive and negative values (Figure 4.12). If we cross that boundary when adding or subtracting, the result is in error.

The reason unsigned arithmetic usually yields the correct signed result is that values are arranged in the same order in each case. It's easier to see that once you notice that the MSB is the only digit to make a negative contribution. When it's 0, signed and unsigned values clearly align with state. When it's 1, values become less negative as the remaining bits count up, meaning these values too align.

An additional aim should be *cheap negation*, but performing a full subtraction seems expensive. There is a little trick which makes it far less so. First, note the definition of "one's complement":

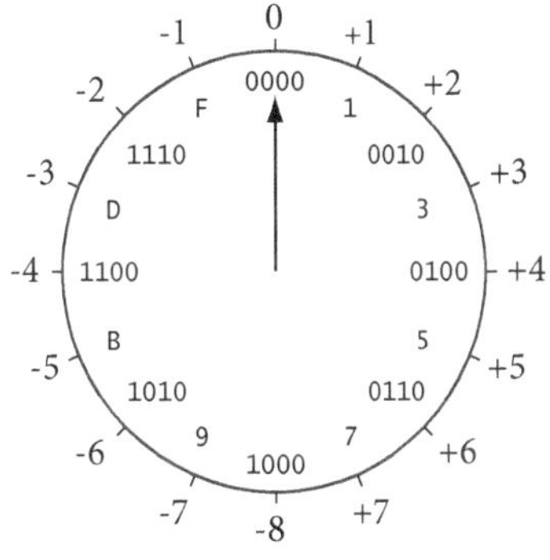

Figure 4.12 Two's-Complement mapping.

$$\alpha^{(1)} = (M-1) - \alpha$$

This amounts to just 'flipping' each bit, since $(M - 1)$ is simply a row of set bits, and turns out to be easily accomplished *within* a register, with no need to move data to and from arithmetic hardware.

If we merge (subtract) the two definitions, we find something very useful:

$$\alpha^{(2)} = \alpha^{(1)} + 1$$

We can negate any value by flipping each bit and then adding one. When we come to build registers, we find that it's possible to perform both operations without moving the data out of its home.

The fact that two's complement requires two-step negation is its only downside.

Communication

Synchrony

There are few things as significant in human history as the birth of telephony. It is hard for us now to perceive hearing the voice of someone far away as the miracle it truly is. Sound itself could be transported some considerable distance by mechanical contrivance alone. "Speaking tubes" had been in use on ships for many a year, but to conquer geographical separation required the use of an 'analogue' — some other physical measure that varies in harmony, and acts as a medium.

A 'transducer' is required at each end. To transmit sound, a microphone and loudspeaker fill that rôle. Either electrical current or potential difference (voltage) can serve as medium. Either way, the physical connection is then simply a cable, comprising two wires. Any such mechanism, complete in every detail and ready for use, via which messages may be conveyed is called a 'channel'.

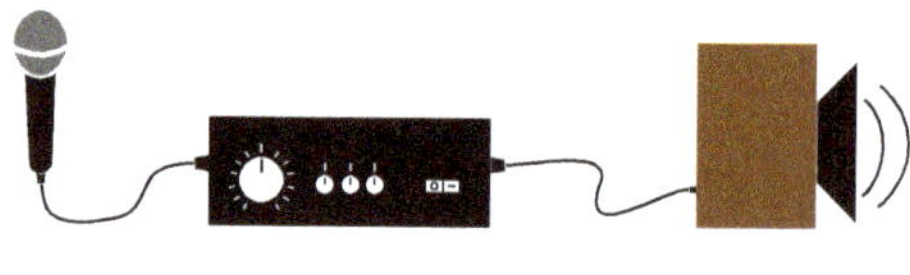

Figure 4.13 Analogue communication.

Microphone, cable and loudspeaker, together implement a channel for the *synchronous* communication of messages. 'Synchronous' literally means sending and receiving a message at the same time. In general, the term applies to any communication where both sender and recipient must wait for each other to become ready, and where the signal or message is passed directly.

Sadly the term is often used loosely, and with a meaning that varies with context.

Wonderful the telephone may be, but it can also be a major cause of interruption and disruption. Sometimes, it is highly preferable to remain uninterrupted for a while — for example, to sleep or finish writing a section of a book — and communicate only when you choose. For some, email recovers some of the peace and control lost with the phone. Mail, whether paper or electronic, constitutes *asynchronous* communication. The sender need not wait for the recipient to become ready. This necessitates a 'buffer' of some kind, where messages may be deposited awaiting collection.

Again, be warned of other (imprecise) uses of the term 'asynchronous'.

Given the ability to conduct synchronous communication, we can always achieve asynchronous communication. However, any asynchronous system can fail: the recipient may arrive early and miss the message, or the sender may overwrite a message before it has been read. Paper mail denies both errors. A recipient knows they are early when they find their mailbox empty, and mail is free to pile up there (to a point). Such is not necessarily the case with electronic systems.

Another example of asynchronous communication is that between lines in a sequential computer program. A value is assigned to a variable, which acts as a buffer. The value is later read and used in an expression, typically on the right-hand side of another assignment. The same variable can be used for many communications; each time, the value is overwritten. It falls to the programmer to ensure that no variable is read before it has been written, and that each is read before it is written again.

Failure to do one or the other is a common programming error.

Asynchronous communication of analogue signals required the perfection of a third electromechanical device — the gramophone (or Dictaphone). In its original form, it transcribed sound into a varying indentation on a cylindrical surface and back again. For the first time, one could record sound and play it back when one liked, provided one could afford both the apparatus and the media.

Synchronization

We have seen how all communication can be categorised as either synchronous or asynchronous, according to whether the message is communicated directly or awaits reception in some buffer. Synchronous communication is the more fundamental, since it is necessary in order for sender and recipient to interact with a buffer. It is also clearly more efficient and secure, though less convenient.

To achieve synchronous communication, some form of *synchronization* is required, to inform each party that the other is ready. Any telephone must have a chime or "ring tone". Similarly, the caller must be able to tell when the receiver has "picked up". Here is something equally fundamental.

There are just three known ways in which synchronization can be achieved. If you come across a mechanism which does not match one of these, apply for your Nobel prize immediately.

They are all familiar from everyday life.

The first, and simplest, is called "common-clock" synchronization. It is nothing more complicated than an agreement to communicate at a specific point in time, perceived via shared access to the eponymous clock. Repeated communications might occur at regular intervals, perhaps upon each *tick* of that clock. Because of its simplicity, this is how nearly all of our digital systems currently operate.

You might say they are 'clockwork'.

There are, however, disadvantages. If this is how people in an office environment worked, their employers would be less than happy. Nothing is achieved between clock 'ticks'. Furthermore, there might be insufficient time to complete many a task. All in all, it's not very efficient, and often impractical.

For small systems — say, an office with only half a dozen people — it's easy to mount a clock on the wall, where they can all see it. But it's not so easy with an entire office block. Indeed, with digital systems, it's becoming a major difficulty, and yet another inefficiency; the same tick-tock signal must reach everywhere in a system which may now comprise billions of component parts.

The second option is a bit better, though more complicated. 'Handshake' synchronization requires two distinct signals, as a handshake needs two hands. Each party uses their own *local* signal to indicate that they are ready to communicate. These are often called 'flags' for the plain and simple reason that brightly coloured flags were indeed used, for example, on a battlefield.

This is as good a time as any to qualify that term 'signal'. Suppose you know in advance what the message is; for example, 'fire!' You just need to know when. No message need be sent; synchronization is all that is needed. This also serves to illustrate another point which arises throughout the science of communication and computation: it is more useful to say the message existed, but was *empty*.

With handshake synchronization, we no longer need to distribute access to a common channel across our entire system. Instead, we need two signals between every pair of communicating components. Communication has become locally, rather than globally, synchronized.

That the number of signals has doubled is not great news. Much worse, the inefficiency inherent in common-clock synchronization survives with handshaking. You now have to wait indefinitely for your partner to become ready, and you have to remain "busy waiting", free to do nothing else.

Handshake synchronization is used in our digital systems, though only between subsystems of significant scale, because of its greater expense. The principal advantage is that it removes the need for all processing to proceed at the same rate. Slower processes can respond in their own time.

The final form of synchronization is optimally efficient, in principle, but is hard to implement. It is far too expensive to use among elementary components, but has been rendered in computer hardware with great success. It's called 'rendezvous'.

Suppose you are a spy, and wish to pass secret information to your handler. Both you and they are very busy people who cannot afford to sit around waiting for the other to get ready. You agree to meet beneath the clock tower at Waterloo Station. The first party to arrive, whoever that may be, leaves a message on a sticky note saying where they are waiting; say, in a café nearby. The second to arrive merely follows the directions on the note and completes the rendezvous.

The essential advantage is that the first party can be busy with things other than waiting.

Signal and noise

Theory and practice are always very different, in any field. Those who had the burden of making telephony work experienced many problems. These can be summarized in two words — *attenuation* and *noise*. Of course, attenuation is the major problem with direct communication, where the medium is air pressure. Even the powerful voice of an opera singer can only carry so far, because the energy employed is dissipated in all directions at once. It is spread over a spherical surface, whose area grows as the square of its radius. The area of the receiving diaphragm is fixed and thus intercepts four times less when the distance from the speaker doubles. Channelling sound along a pipe avoids this problem, though energy will still be given up as the wall of the pipe vibrates in sympathy. However, dissipation is then directly proportional to distance only, and not its square.

The human ear has evolved over millions of years, driven by the need to survive the stealthiest predator. But there are limits to what even evolution can achieve. The smallest variation detectable in the medium concerned defines the sensitivity of the receiver. Sometimes, a certain degree of force must be applied before a membrane will respond at all — a *threshold* must be exceeded. Sensitivity will vary with the signal frequency. Sensors can be damaged by signals of excessive amplitude.[14]

In both artifice and nature, communication is limited by unavoidable vibrations occurring in the environment of the medium, sometimes within the mechanism itself. These vibrations are called 'noise' (Figure 4.14) and typically occur across a broad frequency range. Anyone who can remember the now obsolete gramophone will understand why it's called noise. The act of reading the vinyl record would itself cause damage, adding to the noise encountered next time.

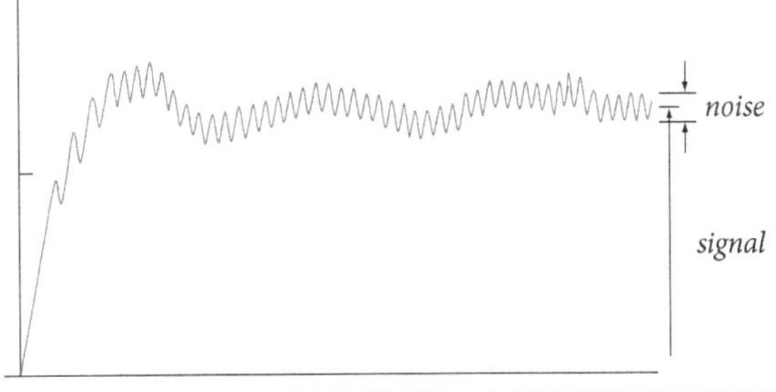

Figure 4.14 An analogue signal and noise.

This is typical of an analogue medium.

Noise also limits the distance over which synchronous analogue communication can be achieved. Repeaters must be employed that amplify and regenerate the signal at intervals. Trans-oceanic telegraph and telephone cables had to site these on the sea bed. When they failed, they had to be retrieved and replaced — not an easy trick. At least transduction to an electrical medium made automated repeating possible. With a voice pipe, and no transduction, the only option was to interpose people along the path to repeat the message.

Analogue communication is inherently limited, over both time and space, because amplification will apply to noise as well as signal, and regeneration will inevitably accumulate errors.

Both nature and technology demand something better.

Sampling and integration

No sensor works instantaneously, for a very good reason. It would have no opportunity to collect any information if it did so. For example, suppose our medium is the force applied via a push-rod. To decide, at the far end, the degree of force applied, it is necessary to permit movement, perhaps the depression of a spring. The more the spring is depressed the greater the force exerted. A measure could be read from a ruler behind. But it takes time for the spring to move. Also, the whole point of the system is that the signal (force) varies with time. If we desire greater precision in our measurement,

14 The author has very low sensitivity to high frequencies in his left ear as a result of once sitting too close to a trumpet.

we must wait a longer time. This means we have a lower temporal 'resolution'. There must always be a trade between information recorded about time and space.

Less uncertainty in one means more in the other.

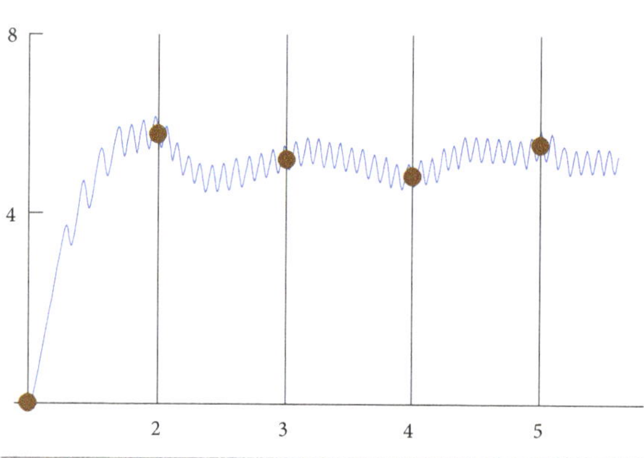

Figure 4.15 Sampling an analogue signal.

Any sensor (microphone, ear, camera, eye *etc.*) can be thought of as taking discrete samples of the medium concerned (Figure 4.15). Apart from the time needed to make a measurement to any given precision, there is also the need to reduce the effect of noise. Some fraction of the inevitable noise in the sensor environment will be added to any signal (the ripple in Figure 4.15). If the sensor *integrates* for a sufficient period, the effect of high-frequency noise should nullify.

Integration appears legitimate once we realize that the signal near the sample point is *correlated*. It can thus make an 'informed' contribution. Signal nearer the point is thus given greater weight; that further removed a lower one.

If a sample is missed then we might interpolate an estimate of its value. CD players use this trick to achieve a smoothly varying output when part of the disc surface is damaged and samples lost. Similarly, a sensor makes use of the signal it detects between samples. In effect, it adds up everything it sees and records a weighted average. The sampling rate and integration time are thus related.

Anyone familiar with electronic technology will be familiar with integration, though they might not realize it. A capacitor accumulating charge is integrating electrical current. The word 'integration' just means adding up, or accumulating. A capacitor simply accumulates charge. To use one as a sensor of electric current, we need only periodically measure the potential difference across it, before draining the charge away. Simplicity begets coarseness; no weighting is applied across the cycle.

We are interested here in *discrete* (sampled) systems — those which communicate by passing discrete messages. However, it is well worth noting that it is possible to integrate a signal and generate a continuous output. Instead of draining the capacitor after reading each sample, we connect a resistance across it so that it drains continuously. One can imagine a balloon with two orifices, one for the signal and one for the drain, which need only be a constricted pipe leading out of the window. Output is the balloon pressure, which could move a dial by some contrivance. An analogue systems engineer would call this a low-pass filter, as any signal (or noise) of high frequency will be attenuated to a much greater degree than those of low frequency. Most electronic sensors are built this way because electrical noise is predominantly of high frequency.

Carrier and bandwidth

Before we leave analogue (continuous) systems for good, there is one other matter worthy of mention — the rate at which we can convey information over an analogue channel. The reason we address this issue is simple and fundamental. Every digital channel is ultimately built on top of an analogue one, usually called the 'carrier'. Should civilization fall and the need thus arise to rebuild our technology, we would have to learn how to build analogue channels first.

A limit to the rate of communication is set by the highest frequency transmissible. Since there is usually little constraint on how low a frequency can be, it is the width of the band of frequencies that matters. Greater 'bandwidth' allows a greater rate of communication.

It should seem reasonable that if more rapid variation can be inferred from a signal then more information is passed per unit time. The highest frequency variation determinable will approximate half the sampling frequency. This is often loosely referred to as the "Nyquist Criterion" after H. Nyquist who published seminal work on telegraphy in the 1920s.

Two insights may help. First, imagine (or draw) a sinusoid through any three points completing its cycle over just two sampling intervals. Given the three points, the exact sinusoid can be recovered. The general equation for a sinusoid has three parameters, initially unknown: amplitude, phase, and frequency. If all three points intersect the same wave then the three parameters can be recovered. Now imagine (or draw) another sinusoid of twice the frequency but still passing through the same three points. It has frequency equal to the sampling rate, completing a cycle within a single period. Such a signal will not be detected. Any such variation is known as an 'alias' of the original signal, since it shares the same signature and is thus indistinguishable. On the other hand, any signal of half the original frequency has a distinct signature and *will* be distinguished.

What emerges from this is that 1) if we wish to convey information at a greater rate then we must use a carrier channel with greater bandwidth,[15] and 2) we must employ a sampling rate of at least twice the highest frequency variation we wish to recognize.

Quantization and digital representation

Communication using analogue signals suffers from the combined effects of attenuation and noise. This can lead to a loss of information at the receiver, where the signal is sampled and interpreted.

---

15 The term 'bandwidth' is now commonly used to denote "maximum rate of data transfer", measured in bits per second. Take care to note that the information transferred (in binary units) may be rather less than the number of binary digits, though it can never be more. For an example, take a picture with the cap on your camera and mail it to a friend.

Suppose we divide a medium (*e.g.* electric current) into numbered bands, and decide into which each sample falls. The sampled signal may now be represented as a stream of digits. This is called 'quantization' (or 'digitization'). Given binary quantization, the signal forms a "bit stream".

As an example, a Compact Disc (CD) stores audio information as a bit stream, recorded as 44,100 16-bit samples per second for each of two stereo channels. A CD player can thus reproduce sound at a frequency up to approximately 22kHz — above the limit of most human ears.

The capacity of any digital channel is dictated by the maximum rate with which one can switch between extreme bands (*e.g.* from the highest to the lowest). This provides us with our first illustration of the primary advantage of binary representation — its inherent efficiency.

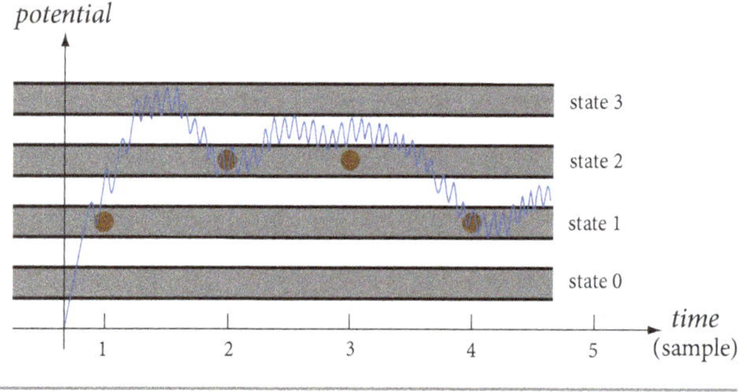

Figure 4.16 Quantization of an analogue signal.

Figure 4.16 shows quaternary (base four) quantization. Four bands require a signal capable of varying three times faster than binary, with two bands, to maintain the same rate of transmission of digits. But each digit carries just twice the information. We could instead send three binary digits. With the same underlying analogue medium, we achieve 50% greater carrying capacity using binary.

By banding states (digits), in this way, it becomes very difficult for noise to cause any error. The second great advantage of digital, over analogue, communication is its immunity to the effects of noise, which may be why nature employs it, with DNA.

Any digital recording benefits from a degree of immunity to degradation over time, similar to that over noise in its creation. With care, a *perfect copy* may be made at any time. By recording some additional information, error correction is even possible, yielding a residual error rate so low it can often be ignored. Only nature betters technology, when copying DNA. Of course, publishing companies and media moguls do not consider this a good idea at all. Witness the row over internet music downloads. But this author is old enough to remember when ordinary folk actually played musical instruments themselves, however imperfectly. Recorded music all but annihilated live music and thus destroyed much pleasure and culture. Perhaps music, film and literature must be free to survive.

There is yet one more big advantage to digital communication. Suppose we wish to improve the resolution of our signal. For example, perhaps we might like high-definition television (HDTV), where each image has a resolution approaching that of a photograph — say, around 3,000 by 2,000 pixels (picture elements) instead of the miserly 800 x 600 of a standard-definition broadcast. With a digital system, all we have to do is transmit more digits to achieve any resolution we like. True, this does require the time or the capacity necessary, but it does not require any major change to the system.

With an analogue system, capacity depends upon the quality (*i.e.* cost) of the components. To improve the system, *every* component may have to change. Evolution is hideously expensive.

As an aside, the change to digitally broadcast television was often accompanied by claims that greater capacity became available. This was only true because analogue systems broadcast each image (frame) continually. A digital television can store images until it needs them allowing the broadcaster to transmit other information. There are two ways in which advantage may be taken of this opportunity: HDTV can bring the quality of the cinema screen into our living room (along with "surround sound"); alternatively, more channels can be broadcast (about thirty in place of one). Sadly, we are mainly getting the extra channels, because this raises advertising (or rental) revenue for the provider. The consumer then suffers from a proportionally lower budget available for TV production, arguably reducing quality. The only valid argument against the introduction of HDTV was the high cost of a worthwhile receiver. However, this soon collapsed. Flat panel displays with the necessary resolution are now standard and the processing and storage required is no longer expensive. High definition has inevitably become the new standard, once the market for additional channels was saturated.

No argument is ever wholly one-sided. Analogue systems had many advocates, for one very good reason: when you digitize, you throw some capacity away. All variation in the medium that takes place within bands is ignored, despite the fact that it too can carry information. There are still some applications that require all the bandwidth that can be mustered, and are thus best achieved with an analogue system. One of which the author is aware is a 'nano-positioning' device, capable of locating a probe with an uncertainty well below the width of an atom. There is one other of which he must not speak. Any instrument operating near the limit of its sensor can ill afford to discard information.

Such applications would gain little and lose much by going digital.

Logic

Propositions and their expression

We all know something of logic since much of it is built into our natural language. That many assertions are either true or false, with no fuzzy half-way point, is part of our everyday understanding of the world, as is the idea that others can sometimes be proved, or denied, with certainty. Our language allows us to associate assertions, via the 'connectives' 'and' and 'or', and to establish dependency, via 'if' or 'while'. For example:

```
It is cloudy and raining.

If it is sunny and warm, I shall play tennis.
```

Of course, there may be more information contained in a sentence, such as intention rather than fact, as in the second above, but the *logical* content is clear. We need to express that, frequently.

Something else we accomplish often is logical *inference* (deduction). For example, the second sentence allows to deduce whether the speaker will play tennis, according to the truth of both premises. Their consequence is called a 'predicate', as opposed to a 'proposition', which lacks any dependency.

Here, the means of inference is implication. There are other ways.

We can form a "truth table" to establish all possible consequences of one or more premises. Because we can recursively compose expressions, taking no more than two at a time, we should begin by tabulating the possible consequences of just one premise (Figure 4.17), and of combining two (Figure 4.18).

x			NOT	
0	0	0	1	1
1	0	1	0	1

Figure 4.17 Possible consequences of a single proposition.

For each premise, there are two possibilities: it may be true or false. We must state all possible consequences in each case. There are thus 2^n mappings from n premises to their consequence. If each row in our table documents a mapping then we shall need 2^n columns to document all of them.

It may seem surprising to learn there are four possible mappings from a single premise. That is because only one is generally useful. The first is called a 'contradiction', where the consequence is always false, regardless of whether the premise is or not, making the latter redundant. In the final column, the premise is again redundant because the consequence is *true*, regardless. Such a mapping is called a 'tautology'. The remaining mappings are termed 'consistencies', or 'contingencies'.

u	v		AND					XOR	OR	NOR						NAND	
0	0	0	0	0	0	0	0	0	0	1	1	1	1	1	1	1	1
0	1	0	0	0	0	1	1	1	1	0	0	0	0	1	1	1	1
1	0	0	0	1	1	0	0	1	1	0	0	1	1	0	0	1	1
1	1	0	1	0	1	0	1	0	1	0	1	0	1	0	1	0	1

Figure 4.18 Possible consequences when combining two propositions.

With *two* premises combined, tautology and contradiction are again possibilities but there are many more consistencies. Some of these appeal to 'natural' logic, or intuition. For example, the second column corresponds to an assertion that *u and v* hold. The fifteenth describes the negation of 'AND', which we refer to as "NOT AND", or just 'NAND'. Similarly, 'OR' and 'NOR' appear as the eighth and ninth columns. "Exclusive OR", or XOR, forms an alternative to the usual inclusive combination.

AND ('conjunction') and OR ('disjunction') are the only 'connectives' appearing in English. Along with negation (NOT), they are sufficient to express *any* logical condition or assertion. For example, exclusive disjunction of two assertions *A* and *B* may be conveyed via "*A* or *B*, but not *A* and *B*".

There is need of no more in any natural language.

Logical calculus and computation

As well as making assertions, through combining and/or denying propositions, it is possible to *evaluate* their truth according to that of their premises. We do this in a way similar to the method used in arithmetic and algebra. It turns out that the rules for combining logical variables are much the same as the ones we use for numerical ones. Things are simpler in logic since a variable can take only two distinct values, denoted by `true` and `false`, or 1 and 0.

When we simply make an assertion, we also declare the truth of each premise. When we cannot rely on our premises, we must carry out a *computation*, based upon their current "truth values". A consequence of any set of premises can always be defined by a truth table, as in Figure 4.19. It may then be expressed as a function of its premises, using the set of elementary *operators* { AND OR NOT }.

u	v	X
0	0	0
0	1	1
1	0	1
1	1	0

Figure 4.19 Definition of function X.

Because { AND OR } obeys the same laws of association and distribution, George Boole proposed representing them by $\{\times\ +\}$. Negation is indicated by a *vinculum* (a bar above a value). Replacing '×' by '.', just as we do in numerical algebra, the function defined in Figure 4.19 may be expressed:

$$X = u \oplus v \quad = \overline{u}.v + u.\overline{v}$$

We arrive at this expression by first noting each row where 1 is desired as output. Then we construct the conjunction which would derive it. Finally, we 'OR' the results (combine in disjunction).

Something profound emerges from this method: *any* logical function may be expressed in a manner requiring just three computational steps: first, we negate certain input variables, leaving the remainder unchanged; then we 'AND', as required; and finally, we 'OR' all the results. Regardless of the number of input variables, a three-layer system will suffice to compute any function.

Recall that only the operators AND, OR and NOT are required.

We have already mentioned George Boole's contribution, in Boolean Algebra. Another legendary British mathematician to leave his mark was Augustus De Morgan, who contributed two laws:

$$\overline{u+v} = \overline{u}.\overline{v} \qquad \overline{u.v} = \overline{u} + \overline{v}$$

which, along with the Laws of Negation and Idempotence:

$$\overline{\overline{u}} = u \qquad u.u = u,\ u+u = u$$

allow a quite astonishing conclusion: { NAND } and { NOR } also each comprise a sufficient set of operators. Hence, we can compute any logical function using a *single* operator (chosen from two options). For example, the function defined in Figure 4.19 can be expressed using only NAND:

$$X = \overline{u}.v + u.\overline{v} \quad = \overline{\overline{\overline{u}.v + u.\overline{v}}} \quad = \overline{\overline{\overline{u}.v}.\overline{u.\overline{v}}} \qquad \overline{u} = \overline{u.u} \qquad \overline{v} = \overline{v.v}$$

Automata

State

For some reason this author cannot understand (and there are many things about our species which defy him) many people seem fascinated by machines which appear to behave like them. While "tin men" have kept Hollywood in business ever since *The Wizard of Oz*, real robots have no reason at all to resemble humans, and every interest in *not* replicating their behaviour. For example, if you're roving around Mars looking for life, getting drunk or setting fire to something in your mouth is not very productive. However, the effort that went in to producing the appearance of life brought reward.

Throughout the 18th and 19th centuries, mechanical 'automata' proliferated. They can still be found decorating church towers and civic buildings right across Europe. A personal favourite is a mechanical wife who appears in pursuit of her mechanical husband, armed with a rolling pin. (This may be the result of having borne witness to such things, growing up opposite a pub.)

State		*Output*
name	*id*	*R A G*
stop	0	1 0 0
pre-go	1	1 1 0
go	2	0 0 1
pre-stop	3	0 1 0

Figure 4.20 Traffic light state and output.

There was a good reason why automata were constructed where they were: there was a clock nearby. Mechanical clocks present an analogue of time, but are capable of digital output also. Their mechanism marks time by repeating a distinct tick, and the inevitable accompanying tock as the pendulum swung back or the 'escapement' resets. Ticks may be counted, and digits advanced.

To design an automaton, you must decide how it moves between distinct 'states'; for example, the rise and fall of that rolling pin. Each state marks three things: what has already occurred, what may happen next and what the audience currently perceives. A more familiar example perhaps is a set of traffic lights, which advances through four states when a vehicle is detected (Figure 4.20).

Given a numerical state identifier, a single register can keep track.

The "state machine"

Our state assignment table goes some way to describing our traffic light automaton but says nothing about how state evolves. For that, we may draw a "state transition graph", from which a "state transition table" may be derived (Figures 4.21 and 4.22).

Clearly, there must be some signal which brings about a move from one state to the next. Hence the need for a clock, whether electronic or mechanical, to generate a sequence of ticks and tocks. All we need do is connect it to the 'control' input of our state register, telling it to change. One possibility is that it causes the register to 'read' a new value to store; another is to increment a count. Either way, we can build a "logic unit" which generates the correct control and data signals from both the current state and the system input (here, the output from the sensor which detects a waiting vehicle).

Remember, we must also ensure the audience sees whatever is appropriate for the current state. With our traffic light system, we need three logical (binary) output signals, to control three lamps. (We may assume that 1 turns the lamp on, and 0 turns it off.) The simplest way is to derive the system output is by using a second logic unit, whose input is simply the content of the state register.

Figure 4.23 depicts a complete system, with its three components.

All we have to do is design appropriate 'logic'.

A "finite-state machine" (FSM or automaton) which generates its output directly from its state alone is called a "Moore machine", after one Edward Forrest Moore, a celebrated computer scientist. But there is one other kind of FSM. It's possible to sometimes "reduce state" (the number of states required, and thus width of the state register) by deriving output from a combination of current state and input. Such a system is called a "Mealy machine", after the similarly celebrated George Mealy.

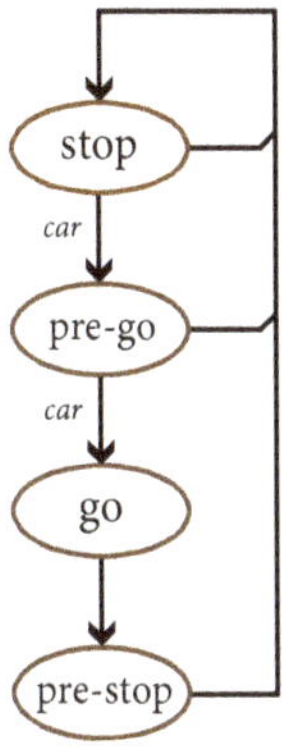

Figure 4.21 State transition graph.

State		*Next state*	
		car	*no car*
stop	0	1	0
pre-go	1	2	0
go	2	3	3
pre-stop	3	0	0

Figure 4.22 State transition table.

Any automaton is a digital system, because it is ascribed distinct states that can be enumerated. It can be fully specified, with absolute precision, and so should be regarded as a formal (mathematical) entity. Because it can precisely describe an animated system, it is of enormous and fundamental importance to computer science. An automaton lies at the heart of any computer.

In fact, it *is* the heart of any computer, as we shall see. But that is only the start.

Every computer program requires an 'algorithm' — a precise procedure. A great many algorithms can be derived and expressed as automata (or "state machines"). Automata enact communication protocols, recognize patterns in text and safely control machinery, among many other things.

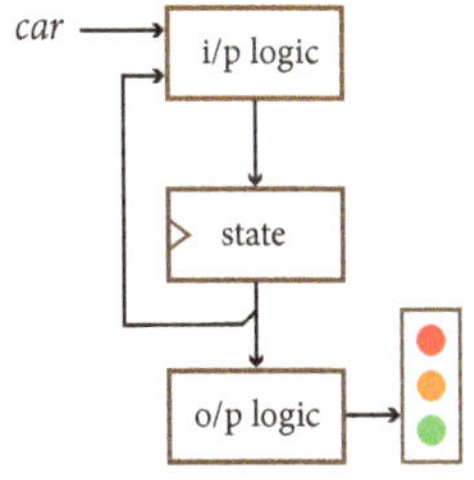

Figure 4.23 State machine.

To build a computer, we must design a system which can effect a sequence of control signals appropriate to carrying out an instruction; for example, to add two numbers together and store the result. Perhaps there are sixteen distinct instructions and thirty-two control signals. Then we need an automaton with four (binary) input signals and five output ones.

Not as easy as it may sound; but we need only do it once, then we can run *any* program.

Application

4.1 Using the Laws of Negation, Idempotence and De Morgan, show that any logical function may be expressed using only the NOR operator.

4.2 Document a Moore automaton that recognizes the word 'else', one letter at a time, beginning with a state/output assignment table. Repeat and adapt the exercise for a Mealy machine.

Which design exhibits fewer states, and why?

4.2 Digital systems

Switches

The normally-open switch

Because we wish to build *digital* systems, we need elementary devices which exhibit multiple distinguishable states that remain stable in the presence of ambient noise. Given the need to build registers with the minimum cost, *binary* (2-state) devices will be preferred. Every component within any system can be ultimately reduced to these devices, which must be capable of communicating with each other.

Besides cost, there are three other things to be kept to a minimum: delay, energy consumption and heat dissipation. Speed and battery life are well-known concerns when choosing a laptop computer, tablet or phone. We are typically less aware of the need to reduce heat dissipation. As the processor in your laptop liberates more heat, either a fan must blow it away or the laptop must rise in temperature so that more heat is radiated. We neither like fan-noise nor being burned.

The fundamental building-block of every digital system is nothing more than a humble switch.

Perhaps the most familiar to us is the light switch. It has two states, corresponding to those of the lamp it controls, allowing or preventing electricity to flow, accordingly.

Each switch will take a little *time*, use a little *energy* and liberate a little *heat* as it operates. Given that a typical laptop processor now comprises more than a billion of them, each operating up to a billion times per second, it is vital to ensure we employ devices that are both fast and efficient.

Like the systems we build, a brain is made up of many interconnected switches, formed from cells called 'neurones'. Only now are we beginning to understand something of neuronal architecture, which is very different from any structure described here. The number of switches in a typical mammalian brain is similar to that in a contemporary computer ($\sim 10^{11}$). However, they operate about a million times more slowly. Yet a brain is capable of feats that we are thus far utterly unable to replicate by programming a computer. Clearly, we have a lot to learn about interconnection.

In both brain and computer, each switch has one stable state, to which it will always return when left alone. Another commonly encountered switch that prefers one particular state is the kind of water tap often used in public amenities. One pushes (applies force) to *close* it, allowing water to pass through. It then automatically opens, after a few seconds. We refer to such a device as "normally open".

Some medium will flow between input and output when any switch is closed. We currently favour electricity because our technology permits us to make very small electrical switches ('transistors') which are extremely fast, consume very little energy and dissipate very little heat. In the past, we have used mechanical switches, permitting rods to move.

In the future, we may well use light, the fastest thing of all.

Electricity is invisible, making exploration of its behaviour all but impossible. However, air behaves in much the same way, yet produces effects that are easily explored. Hence, we shall speak of pneumatic devices here, for purposes of illustration. We can easily imagine connecting a balloon to the input 'port', while holding our hand by the output port to witness the effect of blowing into the 'control' port.

Even the spring could be made from a small balloon, in which case the pressure ('potential') at the control port must become greater than that inside for the switch to close. Just as the pressure in a balloon rises with the air blown in, an electrical 'capacitor' may be charged to raise its potential, which must be similarly overcome.

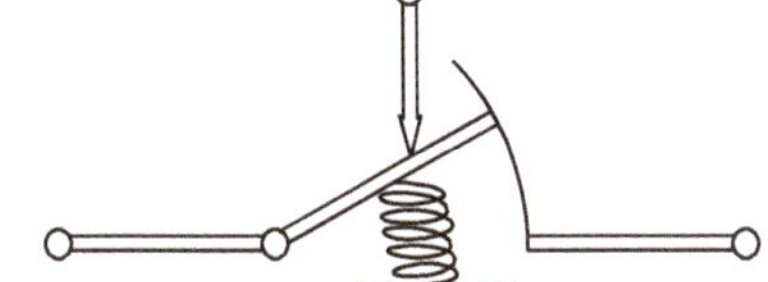

Figure 4.24 A pneumatic normally-open (n-o) switch.

In the world of engineering, the medium used at the control port often differs from that which flows through a switch. A hydraulic device, controlling the flow of a fluid, might well be controlled electrically. Conversely, the light switch on your wall is mechanically controlled. Neither would be of use here. To construct a digital system, we need to connect the output port of one switch to the control one of another, so the medium which flows must also be the one that controls.

The amount of force required to close a normally-open switch depends upon the strength of the spring employed (regardless of its nature). This must be sufficient to ensure it never closes as a result of background noise. Work must be done in closing the switch, and thus energy expended, in direct proportion to the product of force to overcome the spring and 'throw' (lever movement) required.

The electrical analogue of a spring is a capacitor; that of throw, the charge needed to close the switch; that of force, a voltage (potential difference) across the capacitor.

Because a switch possesses two states – open and closed – it may be considered a *logical* device. A logic signal conveys either `false` or `true`. A polarity is conventionally assumed such that a potential adequate to close a 'n-o' switch is associated with `true`, which might be supplied by an air bottle, of the type employed by divers, or perhaps an inflated hot-water bottle. The potential associated with `false` might be provided merely by a vent open to the room, or a hose out the window.

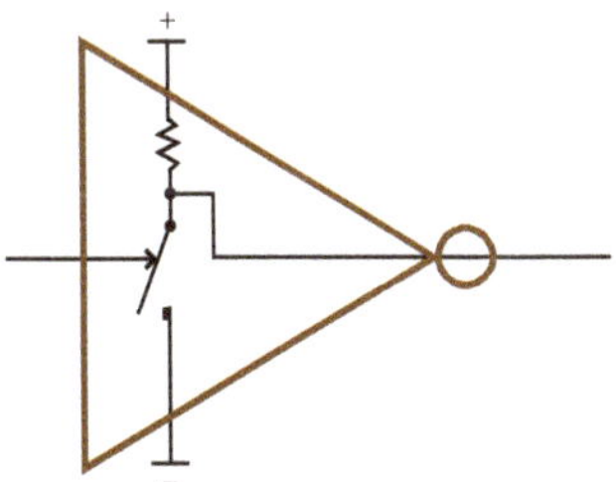

Figure 4.25 shows how a single n-o switch can be used to invert (negate) a logic signal. The enclosing shape represents negation, hiding the mechanism when it is of no concern.

So, why is the resistor necessary?

Figure 4.25 Logical negation via a single n-o switch

Interconnecting switches

When connecting one switch to another, the output port of the first must charge or discharge the control port of the second. This is why it will take a certain amount of time for any effect to appear.

Mass production will lead to complete commonality between switches. They will all be alike. One characteristic will be the resistance R_s to the flow of the medium, whatever it may be, between the input and output ports, when the switch is closed. Another will be the capacitance C_s of the control port, determining how much charging, or discharging, there is to do in order to change state.

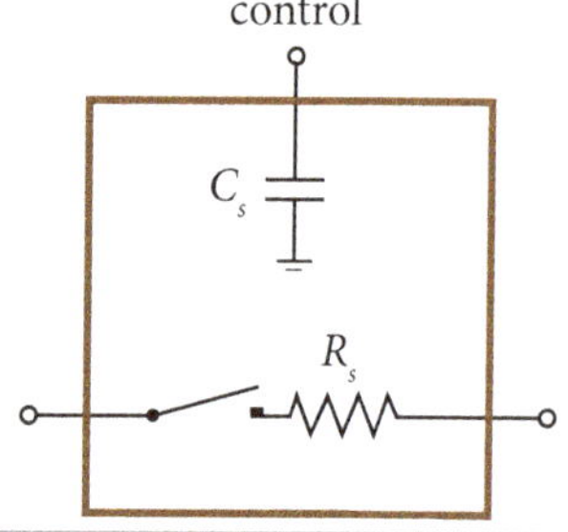

Figure 4.26 Electrical model of any switch.

The speed of operation (opening or closing) varies in direct proportion with both resistance and capacitance:

$$T_s \propto R_s.C_s$$

T_s is often called the "time constant", and is a characteristic of the "fabrication technology" (how the switch is made).

To speed up the effect one switch may have upon another, we might seek to lessen either resistance or capacitance. Recall, however, that capacitance is the analogue of a spring, and must remain sufficient to render operation immune to the effect of ambient noise.

Just as with a river, the rate at which a medium flows is called the current. It will clearly affect the delay in charging the control port of a switch. The larger the potential difference V offered by the power supply, the larger the current I and the faster one switch can affect the state of another, according to Ohm's Law:

$$V = I.R$$

But a higher current means higher power consumption, and thus greater heat dissipation H:

$$H = V.I \quad = \frac{V^2}{R} \quad = I^2.R$$

Note how sensitive heat dissipation is to the voltage (and thus current) employed. It varies as the *square* of either one. This is why the operating voltage of digital systems has reduced over the years, in harmony with our ability to adequately reduce noise or isolate the system from it.

Our entire digital technology comes down to the quality of switch we can make.

We have all become used to the idea of rising processor speeds and the inexorable digitalization of everything once analogue. Music, video, television and photography all now enjoy the advantages of digital systems. All this has been made possible (and affordable) by rising performance and falling cost, enabled solely by a revolution in electronic fabrication technology, which led to a dramatic reduction in the scale of both switch and interconnecting wires (or 'tracks').

Electronic system fabrication capitalizes on solid-state physics to allow "very large system integration" (VLSI) — born originally from the need to cram significant computing ability into the business end of a missile. Competition has driven down the size of the smallest feature the process can resolve. It has done so linearly, with the result that component density has grown *quadratically* (according to the square of resolution). VLSI systems are essentially two-dimensional.

A switch a quarter of the size might operate four times as fast. The spring strength (capacitance) may well change this way. However, conductors also grow narrower and closer to each other. Make a river narrower and you increase its resistance to flow. Worse, every electronic conductor forms an aerial, radiating any change in current and responding to that in its neighbours. The system not only becomes more sensitive to noise, via reduced switch capacitance, but effectively also causes more of it.

We might hope that tracks get shorter but this depends upon architecture. If systems maintain their physical dimensions, as they usually do, tracks will only grow shorter if we adopt an approach whereby components communicate only *locally*, between neighbours. Any global communication, where a component talks to another regardless of its location, will exacerbate the problem.

In short, a *scalable* architecture, ideally suited to exploit the fall in feature size, should involve replication of broadly similar components that communicate locally. In place of one processor, we could instead fabricate more than two thousand of the kind we made a quarter-century ago. If we could get them all working efficiently in parallel, at the same speed, they would yield several hundred times the overall performance. This suggests that a strategy of replication might also have other benefits.

Suffice it to say that the nature of the machine must periodically account for the changing nature of the fundamental (atomic) element from which it is built.

For speed of operation, we need the resistance through the switch R_s to be as low as possible, but this creates a problem when we interconnect them. To illustrate, we shall consider how we might construct a simple one-bit memory, called a 'latch', from just two n-o switches.

Recall that memory is simply the labelling of discrete physical states, and that the cost of memory can be minimized by combining binary devices. We therefore seek to construct a *bistable* device.

Figure 4.27 depicts how this may be designed, and implemented using pneumatic technology. The problem is fairly obvious. If we simply connect the two switches back-to-back, as shown, should one

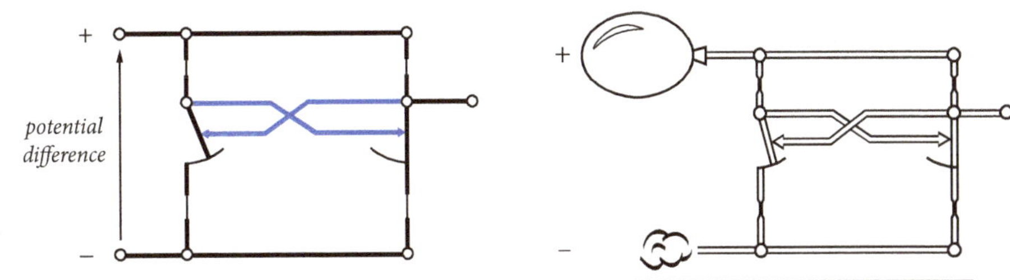

Figure 4.27 A latch made from two n-o switches and its pneumatic implementation.

of them close, a "short circuit" will exist across our power supply. There will be an almighty bang, and our device (and perhaps our hearing) may be damaged.

The solution is to add significant resistance either side of each switch. If we use plastic hose to connect things up in our pneumatic implementation, bulldog clips will suffice.

But there is a price to pay: operation will be slower, and will dissipate more heat.

Consider what happens when we first connect our power supply. Both switches are initially open. Air will bleed past both upper resistors to charge the control port of each switch. Should both switches require exactly the same time to close then both will immediately spring open again as the charge leaks through the upper flow port of its partner. The result will be *oscillation* — both switches will repeatedly open and close. However, it is very unlikely that two components will ever be identical.

Even the slightest difference between switches, or their connections, will allow one or other to close first. A *race* will occur. The winner is generally unpredictable, because we are unlikely to know enough about our components. But the switch that closes first will decide the stable state the system will then adopt. There are two such states, which might be labelled `0` and `1`.

Race and oscillation are the two possible results of 'feedback', where output is allowed to affect input. A race is indicative of two or more stable states, oscillation of one or more unstable ones.

The normally-closed switch and complementary pair

It turns out to be typically as easy to make a normally-*closed* (n-c) switch as a normally-open (n-o) one. In our mechanical picture (Figure 4.24), all we need do is move the spring to the other side of the lever. Fortunately, it is as simple with the somewhat faster electronic switches we currently use.[16]

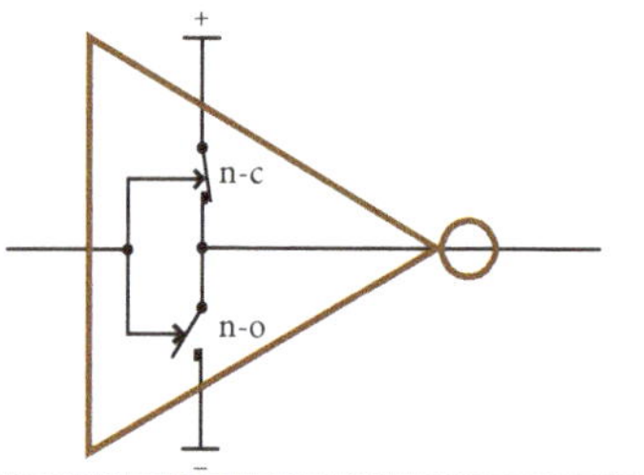

Figure 4.28 Negation by complementary pair.

By using a *complementary pair* of switches, one normally open, the other normally closed, it becomes possible to design every component we require *without any additional resistance* (or capacitance). As a result, each will operate with a speed, energy consumption and heat dissipation limited only by the quality of the switches themselves. All we need to do is connect them as efficiently as possible.

The simplest illustration is a revised design for logical negation (Figure 4.28). When the input is asserted (*i.e.* `true`), the n-c switch closes and the n-o one opens, leaving the output `false`.

16 The best mechanical switches might close in a millisecond, electronic ones in less than a nanosecond — at least a *million* times faster.

A little thought should reveal the one and only condition which must govern each and every complementary pair: *the normally-closed switch must open before the normally-open one closes.* Otherwise, we shall again hear a bang that indicates a short circuit, fused device and empty power supply. Revising the design for our latch is just as simple, and once more removes the need for added resistance (Figure 4.29).

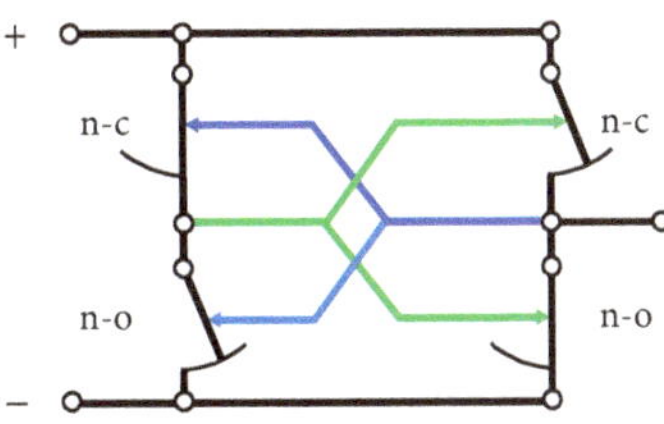

Figure 4.29 Latch by complementary pair.

Designing circuits with resistance and capacitance is the province of the electronic engineer. It would demand knowledge of electrical theory and the practice of electronic design. But we have just seen that we can build networks of complementary switch pairs, and thus any digital system we like, with no such requirement. There is a clear divide between the study of electronics, or indeed any fabrication technology, and that of digital systems.

All we need to understand is how to interconnect our switches to make useful components.

Gates

Nature and symbols

It turns out that arithmetic can be accomplished via logical functions: XOR corresponds with binary addition; AND will compute a 'carry'. So it should come as no surprise that we set out to build a computer from devices which implement elementary logical operations. These are called 'gates'.

From our brief introduction to logic in the previous section, we can identify the gates required with any sufficiency set; for example, { AND OR NOT }. We saw how any logical function can be implemented by no more than three layers of gates connected one after the other, *with no feedback*. The first layer comprises negation (NOT), the second conjunction (AND) and the last disjunction (OR).[17]

There is a long tradition in the graphical design and recording of such systems. Lines depict the connection of the output of one gate to the input of another. The icons that represent { AND OR NOT } are shown in Figure 4.30. The 'bubble' on the nose of NOT may be added to any input or output of an AND or OR to indicate negation, removing the need to draw an explicit gate for the purpose.

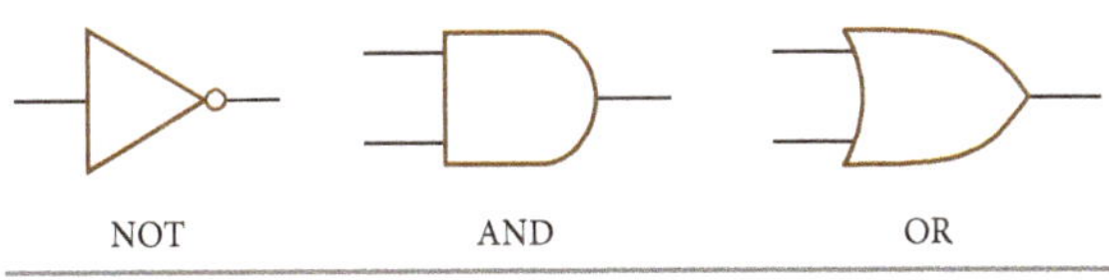

Figure 4.30 Graphical depiction of logic gates.

It should thus be clear how to draw NAND and NOR gates.

17 It is perfectly possible to implement the same function using OR in the second layer and AND in the third. You can demonstrate this either via algebra or directly from the truth table by considering each row with output 0, instead of 1.

Recall that De Morgan's Laws allow any three-layer {AND OR NOT} network to be built using either NAND or NOR gates alone. A graphical demonstration of this is left as an exercise (see below).

We have assumed thus far that all of the binary (two-input) gates can be extended to have an arbitrary number of inputs. It's certainly straightforward to extend their *definition*, via truth table, but a little more effort is required to make sure that two two-input gates yield the same function as a single three-input one, for example. We must be careful.

Gates and canals

Logical AND and OR are clearly special in that they combine (along with negation) to afford the expression of any logical function. No doubt for this reason, natural languages, like English, evolved their inclusion long before this mathematical nicety was discovered. But there is more.

They are each capable of 'gating' a logical signal, and complement each other.

Having found that both may be extended to a function with an arbitrary number of inputs, we now note that any one of them may be used to "switch off" output. In the case of AND, our 'control' input must be 1 to allow the other one to "pass through". 0 will turn output 'off', effectively blocking its path. OR offers corresponding control, but with a 1 blocking the path, and 0 allowing the other input through (Figure 4.31).

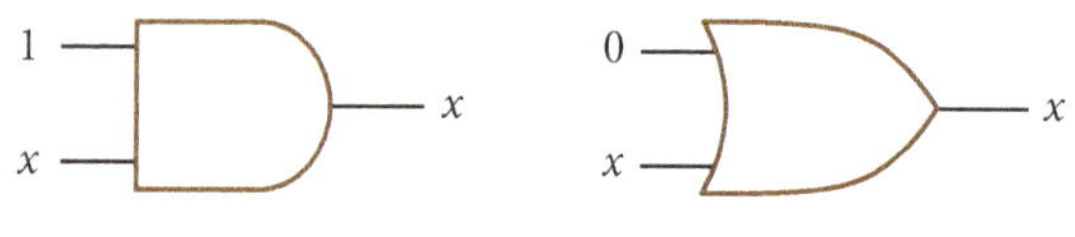

Figure 4.31 Gating via AND and OR.

Either device can thus be understood to act as a 'gate' allowing or denying passage.[18]

But we're not yet done.

Our 'control' input may instead be regarded as one end of a *canal*, the other being the output. Should the right logic value arrive, it will "sail through", like a barge on the canal, regardless of all other input. That value is 0 with AND, and 1 with OR — the complement of the one that shuts the gate.

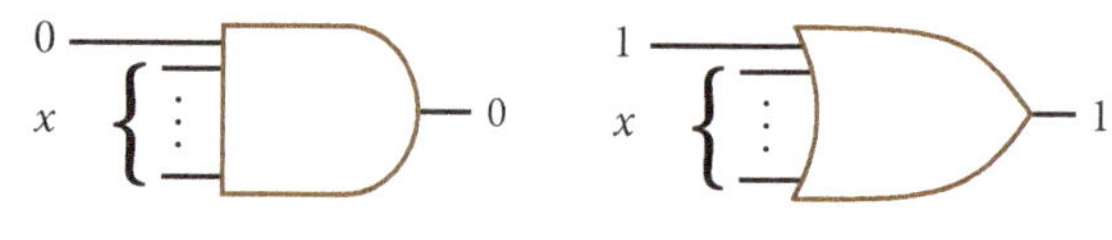

Figure 4.32 'Canalysing' AND and OR.

Other logical operations possess a similar property, where a 1 or 0 is *likely*, but not assured, to sail through. Even logic can deal in uncertainty, and in degree, rather than extremes.

18 The author does not know whether this property is the origin of the term 'gate'.

Implementation

Switches combine intuitively to yield logical function. Two n-o ones in series will yield an AND gate; two in parallel, an OR one. Respectively, output will be asserted if *u* and/or *v* are (Figure 4.33).

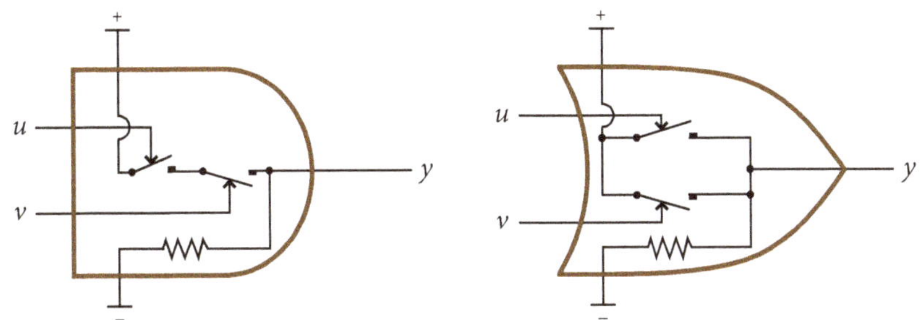

Figure 4.33 AND and OR gates, fabricated from n-o switches.

However, we should be far more interested in how to make either NAND or NOR, since either is sufficient on its own to make any digital system, and it is cheaper to mass produce a single type of component. We have also seen that using complementary pairs of switches is more efficient and leads to faster devices. Figure 4.34 shows how NAND and NOR functions are achieved.

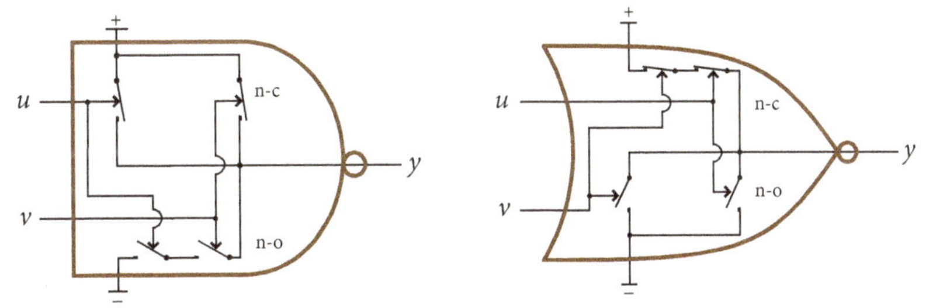

Figure 4.34 NAND and NOR gates, fabricated from complementary pairs.

Recall that the one required condition for any complementary pair is that the n-c switch always opens before the n-o one closes, to deny any short circuit.

To keep things simpler, from here on we shall use only NAND gates, though design may still originate with { AND OR NOT }, derived directly from a truth table, and XOR is occasionally of value.

Application

4.3 Show how De Morgan's Laws may be stated graphically.

Show graphically how your answer may be used to convert a three-layer NOT-AND-OR gate network to one comprising NAND alone, and a NOT-OR-AND network to one comprising NOR alone.

4.4 By adding columns to a truth-table, show that a three-input AND gate does indeed implement the same function as two two-input devices. Do the same for OR. Now try it with XOR.

As with AND and OR, you must first surmise what a three-input XOR gate might do.

4.5 How might a XOR gate be fabricated from complementary pairs of n-c/n-o switches?

Clue: Experiment with one input controlling both switches, while the other feeds through them. It's legitimate to employ NOT (or other) gates in the composition of another type.

Registers

Latch

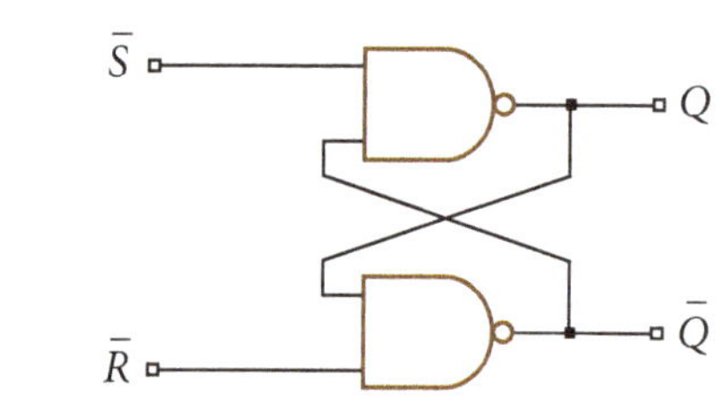

Figure 4.35 A *RS* latch from NAND gates.

While a bistable device to 'latch' a single bit can be made directly from complementary pairs, as shown in Figure 4.29, it will not be easy to use. For example, if we keep output and input separate, they will always represent the inverse of each other. Partly for this reason, and because we wish to build everything from standard components, we choose to make a one-bit latch from gates.

Figure 4.35 shows how two NAND gates may be connected, with mutual feedback, to create a bistable. Like the earlier one, this bistable suffers a race condition, when both inputs are 0. Otherwise, it's well behaved, and can be 'set' via $\overline{S} = 0$ and 'reset' via $\overline{R} = 0$, with the alternate input 1.

Outputs Q and $\overline{Q}$ will complement each other, except when $\overline{S}\,\overline{R} = 0\,0$.

The device has an interesting property in that it will retain its previous state when $S\,R = 11$.

A problem with the latch of Figure 4.35 is that we cannot simply connect a data signal. We have somehow to arrive at $\overline{S}$ and $\overline{R}$. The solution is to add another NAND gate, to invert our data signal D.

The additional gate affords a bonus: we can now "turn off" the device, using the spare input. Because it "turns on" the D input when asserted, convention bestows the name 'enable', or '*EN*', for short.

The complete D latch is shown in Figure 4.36.

Another problem remains: our latch is 'transparent', in that the input is 'visible' from the output. What we mean by this is that any change on the input will almost immediately appear as output. If we wish to chain devices together, to allow data to pass through a system, we need something more.

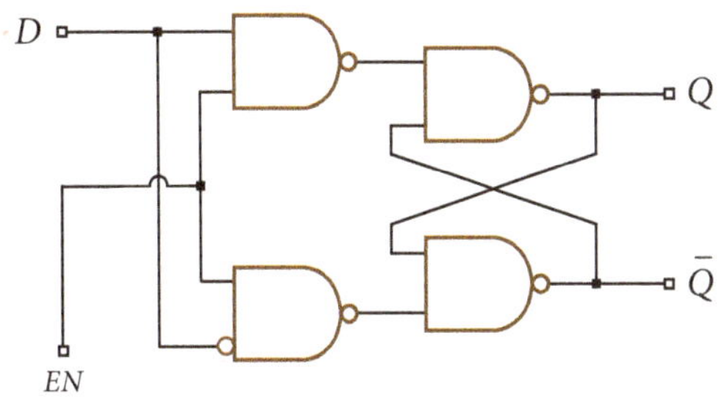

Figure 4.36 The (transparent) *D* latch.

Flip-flop

For one memory 'cell' to communicate with another requires each to possess *two* latches: one to register a datum for output, the other to record fresh input. If synchronous transfer is to take place on the tick of a common clock then we shall also need a 'tock' on which to move input to output. A "two-phase" clock is one which can provide both tick and tock. This need be nothing more than a wire on which the logic value changes from 0 to 1 and back, repeatedly. It can be generated by an unstable system with feedback, such as the one depicted in Figure 4.37.

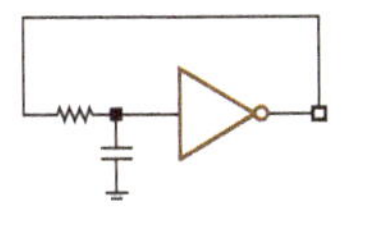

Figure 4.37 A digital clock.

The 'inverter' charges (or discharges) a capacitance via a resistance. Once charged (discharged), the output will invert. The rate at which the clock ticks depends on how great both the resistance and capacitance are. System speed is dictated by the slowest 'clocked' component.

Figure 4.38 The (opaque) *D* flip-flop and its timing.

Figure 4.38 depicts our two-latch device, which is called a 'flip-flop' because of its two-phase action, along with a diagram showing its operation as the clock signal changes. Note how the "no-man's land" between the designated voltage range for each logic value is exploited to isolate the first latch from the second. The second one can only accept input *after* the first has been disabled.

Unlike a latch, a flip-flop is not transparent; input can change, but output will not follow until the next "clock cycle" is complete. The clocked "double buffer" ensures this.

A data register is just a collection of *w* flip-flops, where *w* is the 'word' width required. For example, a 64-bit register is simply an aggregate of sixty-four (1-bit) flip-flops, with no interaction between.

We can add an extra input gate to each flip-flop to allow both *clock* and *enable* register control.

Register transfer and register transform

Armed with a digital two-phase (tick/tock) clock and a number of registers, each comprising *w* double-buffer flip-flops, we can build systems where data may be transferred from one to another. All we need is a data 'bus' (bundle of pipes, wires or whatever) and a little plumbing, as in Figure 4.39.

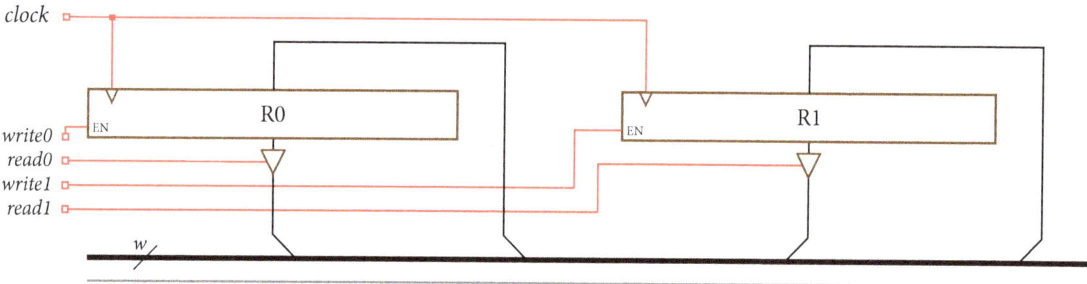

Figure 4.39 Two registers able to transfer data in either direction.

Data is allowed out of each register via a 'crossbar' switch, controlled by a dedicated signal.[19] A row of *w* n-o switches, sharing a common control signal, will suffice for this purpose. A second *enable* control input, common to all flip-flops, decides whether data is allowed *in* to a register.

A certain "control word" will cause a transfer of data from *R0* to *R1* in the ensuing clock cycle, which we write as *R0* → *R1*. Another will cause the transfer *R1* → *R0*. Unlike a data word, the elements of a control word have no intrinsic order, though it is convenient to apply one just for consistency.

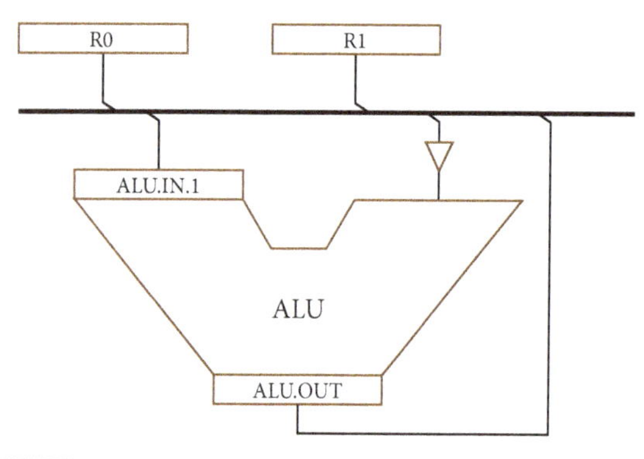

Figure 4.40 Register transform via an ALU.

You may well ask what use all this might be.

Suppose one of our registers forms an input to a device capable of arithmetic, and another latches its output. Perhaps two more control signals decide its precise function. With the correct sequence of control words, we can accomplish an arithmetic operation on the content of any pair of registers. By executing a series of such operations, we could compute any function of the data with which we begin.

Gates are combined to produce 'logic units' which transform the content of one register into that of another. Register transform and register transfer combine to offer the potential for a universal computer.

19 The term 'crossbar' comes from the railways, where a long iron bar was sometimes used to control the setting of multiple points at the same time. A single action could then dictate the path to which two or more tracks were connected.

Register addressing

We have seen how a register can be controlled by two control signals: *read* and *write*. One fairly obvious problem is the need to ensure they are never both asserted at the same time. While nothing will catch fire, nothing will be accomplished either; the register will simply read its own output.

Much more serious problems emerge when we have a large number of registers. First, we really do need to take care that only one read signal is asserted at any one time (tock). When the output of two or more registers is connected to the same bus, there may well be smoke and flames. More likely, with contemporary electronic implementation, components will simply fail and become dysfunctional.

Such a situation is called "bus contention".

There is also the "interconnection problem".

Suppose we have a million registers, which would make up a very small memory for a contemporary computer. To connect anything to such an array would require two million conductors (of whatever medium is employed). Not only would they prove expensive to construct, they would also consume much power and liberate much heat. The matter is made worse by the tendency of memory to grow exponentially over time. Computers have moved from kilobytes to megabytes, and now gigabytes.

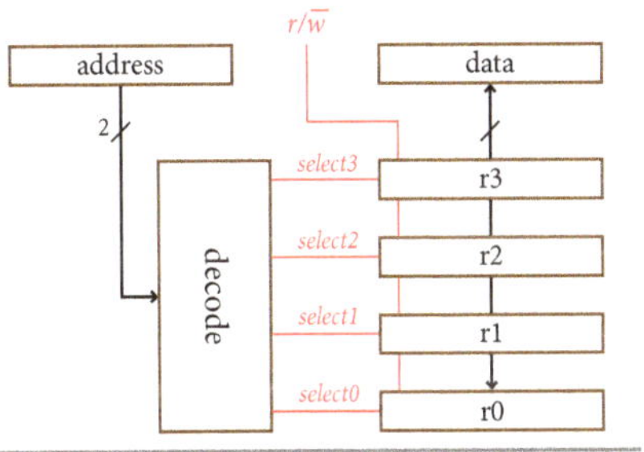

Figure 4.41 Memory addressing.

The solution is to enumerate registers and encode the identity of the one required as a unique number. By comparison with identifying a house on a street, this is referred to as its 'address'. As long as only one register is being read or written at any one time, a single *read/$\overline{write}$* signal need accompany the address. (When asserted, the designated register is read; when denied, it is written.)

Of course, addressing only identifies a single 'target' register. Its partner in the transfer may be thought of as lying at the end of the bus, within the device making use of the memory.

To select one register from n we previously needed $2n$ conductors (control signals). By addressing our register in binary, we now need only $\log_2 n$. This allows memory to grow exponentially without becoming too expensive, power-hungry or hot. A million registers will require just twenty-one conductors (including one for *read/$\overline{write}$*) instead of two million, and a billion just thirty-one.

Application

4.6 Show how a NAND gate can facilitate an extra 'enable' control input to a *D* flip-flop.

4.7 By considering Figures 4.39 and 4.40, and enumerating all control signals required, work out the sequence of control words necessary to add the contents of *R0* to *R1*, leaving the result in *R1*.

What assumption must you make, with regard to timing?

Arithmetic logic

Arithmetic negation

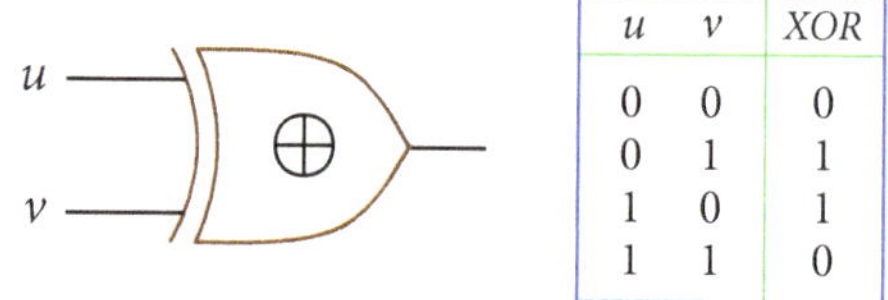

u	v	*XOR*
0	0	0
0	1	1
1	0	1
1	1	0

Figure 4.42 The exclusive-or (XOR) gate.

For arithmetic logic, we make extensive use of the two-input XOR gate (Figure 4.42), which may also be fabricated from complementary pairs of n-o/n-c switches, like NAND and NOR. The most obvious reason, to which we shall return, is that its function is exactly equivalent to summing two bits (binary digits). Perhaps less apparent is that it can effect inversion of one input, according to the other.

Suppose u carries the bit we may wish to invert. When $v = 0$, the output follows u, but when $v = 1$, it follows the inverse of u. We can say that the gate inverts u, according to v.

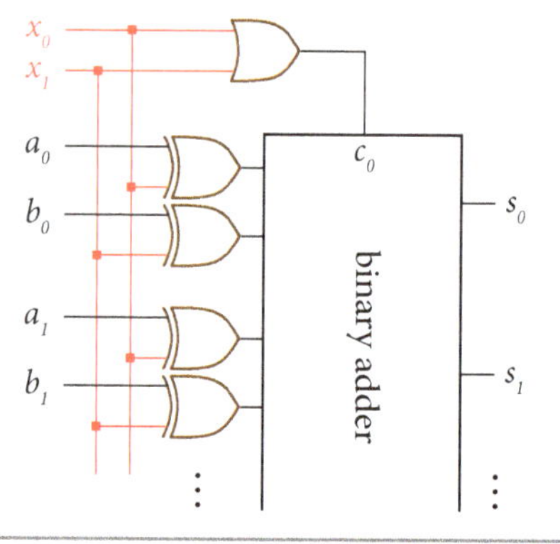

Figure 4.43 Arithmetic negation.

Remember that, using two's-complement representation, the sign of an integer is inverted by taking the ones complement (inverting each bit) of the register content α and adding one:

$$\alpha^{(2)} = \alpha^{(1)} + 1$$

We can accomplish one's complement using one XOR gate for each bit, a_i and b_i, of each input word, A and B, as shown in Figure 4.41. A control signal, c_0 or c_1, commands the negation of each input word. Adding one can be achieved via a "carry in" to the least-significant bit in the addition, formed from the OR of c_0 and c_1.

Hence, according to the state of c_0 and c_1, we can compute:

$x_0x_1 = 00$: $A + B$ $\quad$ $x_0x_1 = 01$: $A - B$

$x_0x_1 = 10$: $B - A$ $\quad$ $x_0x_1 = 11$: n/a.

(We must ensure that we never assert both control signals at the same time.)

Binary addition

A logic unit that adds together two input bits and an incoming carry, to produce both sum and outgoing carry, is called a "full adder". We might achieve this with two XOR gates, three AND and one OR:

$$s_i = a_i \oplus b_i \oplus c_i$$

$$c_{i+1} = a_i.b_i + a_i.c_i + b_i.c_i$$

It's important to remember that we can always implement a dedicated design, beginning with a truth table (with three left-hand columns and two right-hand ones), which would comprise no more than three layers. However, intuition suggests a design with just two layers.

We might simply "daisy chain" w full adders, with the outgoing carry from each connected as the incoming one for the next. A little reflection quickly reveals the flaw in this approach. If w is small then fine, but if it's sixty-four, as it is in a typical contemporary personal computer or tablet, then we will wait sixty-four times longer than necessary for the result. Given both input words, and any composite "carry in", we can design a system which yields the sum after just three "gate delays".

The so-called "ripple-through carry" adder is not an acceptable solution.

Let's return to the mathematics (as any good engineer should, when necessary).

Each equation forms what is known as a "recurrence relation", where c_{i+1} is determined, in part, by c_i. From the truth table for XOR, we can also see that there is a common element between them:

$$s_i = p_i \oplus c_i \qquad \text{where} \qquad p_i = a_i \oplus b_i$$

$$c_{i+1} = g_i + p_i.c_i \qquad \text{where} \qquad g_i = a_i.b_i$$

The p_i and g_i are commonly called 'propagate' and 'generate' variables, for reasons that are apparent when we consider how each successive carry is computed:

$$c_1 = g_0 + p_0.c_0$$

$$c_2 = g_1 + p_1.c_1 \quad = g_1 + p_1.g_0 + p_1.p_0.c_0$$

$$c_3 = g_2 + p_2.c_2 \quad = g_2 + p_2.g_1 + p_2.p_1.g_0 + p_2.p_1.p_0.c_0$$

$$\ldots$$

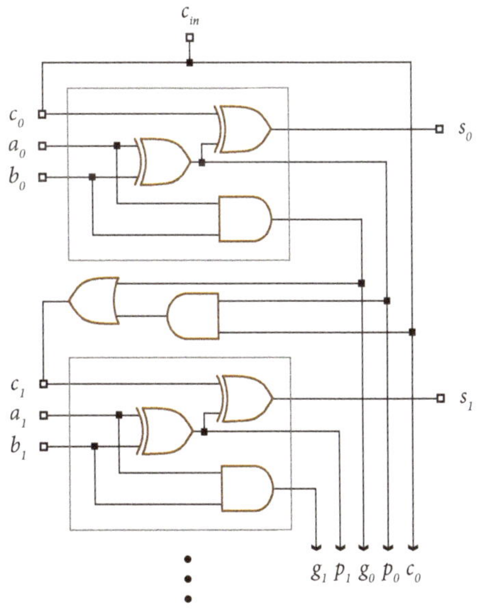

Figure 4.44 First two units of arithmetic logic.

The upshot is that, instead of a full adder for each bit addition, we employ a unit that computes p_i, g_i and s_i, followed by two-layer logic which varies in extent with i, as in Figure 4.42.

Speed of computation is a little worse than ideal, at four gate delays, but far better then ripple.

Application

4.8 Given the first expression for c_{i+1} above, along with the definition of XOR, derive the second one.

Clue: Make use of the Law of Absorption, which states that $A.B + A.B.C = A.B$, which you might easily prove using a truth table.

4.9 An arithmetic-logic unit (ALU) is typically expected to provide 'bit-wise' logic operations as well as arithmetic ones. Show how bit-wise AND and OR may be obtained from the system described.

4.10 Show how a binary memory address may be decoded to enable the chosen memory register.

4.11 Show the logic required to derive register *read* and *write* control signals from *select* (output by the address decoder, specific to each register) and $read/\overline{write}$.

4.3 The digital computer

Essentials

Organization

Within any computer, we have three functions to perform: memory, processing and 'i/o' — short for 'input/output', *i.e.* communication with the outside world, without which the machine would be useless. We need memory for both a program, which defines the processing, and the data, which describes the information necessary, affected or produced.

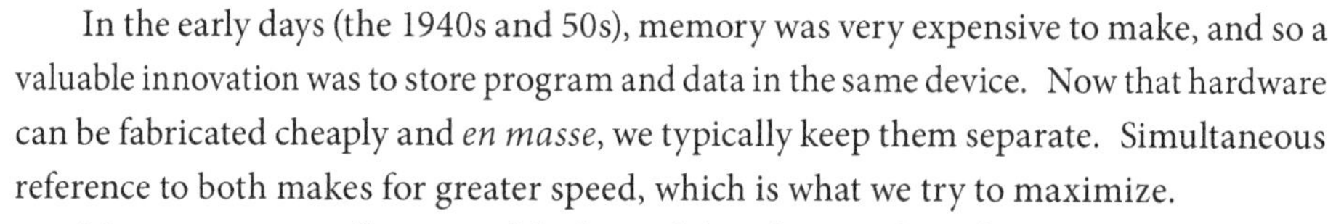

In the early days (the 1940s and 50s), memory was very expensive to make, and so a valuable innovation was to store program and data in the same device. Now that hardware can be fabricated cheaply and *en masse*, we typically keep them separate. Simultaneous reference to both makes for greater speed, which is what we try to maximize.

Memory accounts for most of the kit and thus the cost these days, largely because we have learned to keep the processor simple, but is essentially composed of a very large array of identical registers, which can in turn be broken down into flip-flops and switches.

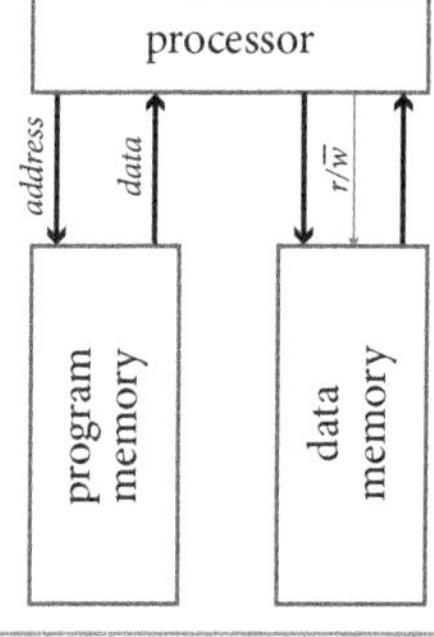

Figure 4.45 Computer organization.

We will describe the processor later, which leaves us with i/o, *i.e.* external communication.

As shown in Figure 4.45, the processor itself already has the means to input and output data. All that remains is to find a way to route it to and from the outside world. Precisely how this is accomplished depends very much on the external hardware concerned. There isn't space here to address how keyboards, displays, disc drives and network adaptors, for example, work, but we can talk about how communication with such devices appears.

A typical device adaptor requires a 'port' where data can arrive, another where it leaves and a number of switches and indicators which control and monitor its configuration. The latter can be contrived to appear as bits within a 'control' register. The two ports are also just registers. All three are 'mapped' within the data memory. By this we mean that they are assigned fixed addresses, allowing them to be read and written by the processor, allowing a program to perform i/o.

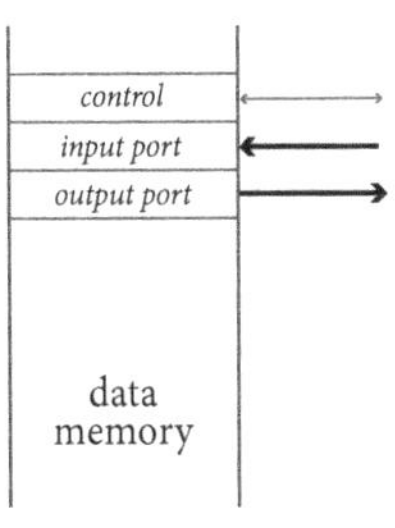

Figure 4.46 Memory-mapped i/o.

Every external device with which the computer communicates will possess an adaptor with control register and ports mapped into data memory. Subroutines that manage them form "device drivers".

While it is possible for the processor itself to enact each external communication, it is far from efficient. The first problem would be knowing when data has arrived, or departed. The processor could repeatedly 'poll' the input port until it finds something new. It would be straightforward to delay a second arrival until the port has been read, leading to a form of handshake synchronization. Departure could update a flag in the control register, which the processor must poll instead.

Polling obviously keeps the processor distracted from the task in hand, and is thus costly.

The traditional resolution of this problem is to have the external device *interrupt* the processor when a communication has been made, allowing it to read/write another datum from/to the port concerned. In response, the processor must execute an appropriate "interrupt service routine" and complete the handshake with the device by returning an "interrupt acknowledgement".

When requesting the interruption, the device must also identify itself via an "interrupt vector", which is normally just a number, encoded on a small bundle of control signals. This can be used to locate the correct service routine simply by adding a base value and looking up its starting address in an "interrupt despatch table". Along with the routines themselves, this forms the core of a typical contemporary "operating system".

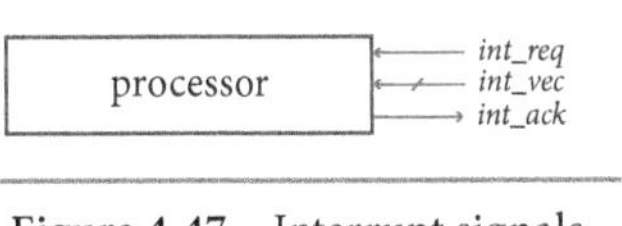

Figure 4.47 Interrupt signals.

The second problem with the processor itself performing external communication is the need to read/write every word of the message and then write/read it to/from where it belongs. This not only doubles the communication with the data memory, it also ties down both processor and memory.

This time the solution is to allow the external device "direct memory access" (DMA), in the same fashion as the processor, instead of the two data ports. DMA requires a mechanism for sharing the 'bus' (address, data and control signals) between devices. Simple "bus arbitration" requires each competing device to signal "bus request" and await a "bus grant" to complete a handshake. Given that the processor usually needs the highest priority, it should come as no surprise that it is typically "bus arbiter". This makes it into a "central processing unit" (CPU), although it was pretty 'central' before.

It is important to point out that there is no intrinsic reason why multiple processors should not share memory, with just one 'central' in the sense of bus arbitration. However, as hardware has evolved over the last half century, processors have gained speed faster than memory. Because of this, *distributed* memory, where each processor possesses its own, would seem to offer a brighter future.

Such a machine requires a communication network and 'multiprogramming'.

Instruction set

The program stored in its own dedicated memory consists of a sequence of 'instructions', each one encoded as a small set of numbers, indexing the:

— operation — the "operation code", or 'opcode'
— operand location — the "addressing mode" for the operand
— operand address — identifying the register containing the operand.

There are two principal locations for an operand: a small "register file", inside the processor, and 'main' data memory, beyond it. The address may then be simply an index into one or the other. However, there are other useful addressing modes. For example, when we wish to refer to an element in an array, stored in sequence within main memory, we must also index that element. A 'base' address might be stored in an internal register, to which that index must be added before reference can be made.

There may be one, two or three operands described, depending upon both the instruction concerned and the architecture of the machine. Most instructions act upon data; for example, `add` will (usually) act upon two operands and yield one result. Since the register addressed for the result is updated, it is also considered an operand, making three in total: two 'read-only', one 'write-only'. A 'three-operand' architecture will detail all three in each instruction, with obvious disadvantages: calculation of the 'effective' address for all may be required, and the program will be laterally bloated.

The "instruction format" will be rather wide, causing the program to require more space.

Effective address computation will mean that some instructions require more time than others.

'Time/space' considerations affect the design of both software (the 'algorithm' employed) and hardware (its architecture). There is always a 'trade-off' between the two, but we prefer to waste neither. Now that we often wish to transmit a program over the internet, saving space can also save time.

A two-operand architecture can be realised by using one for both input and output (result). One operand remains read-only, while the other becomes 'read-modify-write'. Early machines achieved a one-operand architecture by dedicating an internal register for the second rôle. Its use in summing a list of numbers gave rise to the name 'accumulator'.

Clearly, a zero-operand architecture will offer the most compact code. To accomplish this, we need to make use of a brilliant idea whose origins lie in the development of calculators: a 'stack'.

Many a busy restaurant makes use of a spring loaded stack of plates, which always leaves a few above the worktop. As you *push* cleaned plates on, it sinks; as you *pop* them off, it rises. The same idea can be used for manipulating data in memory. We need only keep a record of the address of the current 'top-of-stack' (TOS), usually within an internal register for speed and ease of reference.

Our `add` instruction may now assume that its two operands are ready to be popped off the TOS, and the result pushed on. A series of arithmetic operations may then proceed sequentially, each communicating with the next via the stack. One can even minimize intervention by preloading operands referred to within an expression, prior to evaluation, in the order required.

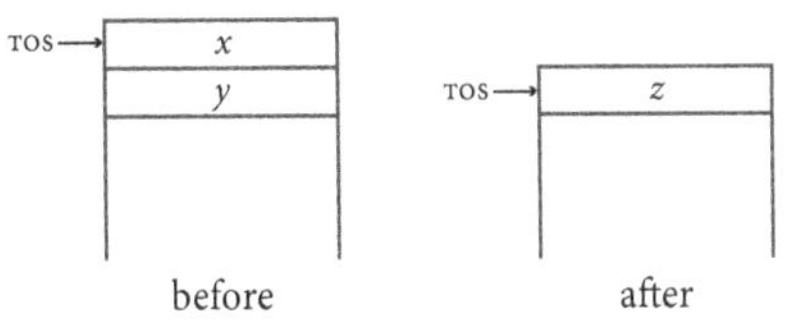

Figure 4.48 Use of stack for an addition

Expression evaluation would then be close to optimal since almost no work is performed that is not intrinsic to the problem. But we need to remember that access to main memory is slow, while that to an internal register is fast. As we mentioned earlier, the difference in speed has been growing for the whole history of computing and is now two whole orders of magnitude.

Fortunately, there is a solution. It relies upon the fact that, while it is dynamic in nature, the stack only grows or shrinks by an item or two. As one instruction pushes something on, the next pops it off. All we need to do is 'cache' the topmost part of the stack inside the processor. As long as that cache is large enough, only rarely will we need to pause execution and refer to main memory.

From experience, we can judge how large the cache needs to be to evaluate most expressions without reference to main data memory. Program 'execution' then reduces to a sequence of efficient load/evaluate/store cycles. Any machine so designed is thus said to possess a 'load/store' architecture.

We have thus far discovered two categories we need within our "instruction set":

— load/store   to move data between the processor and main memory
— operators   to act upon data (`add`, for example).

If we adopt a stack architecture then load and store reduce to `push` and `pop`.

We will need an appropriate set of operators for each data type. For example, we shall need { `add` `sub` `mul` `div` } for each numeric type, including (unsigned) `natural` and (signed) `integer`. It would be better to give them distinct names, perhaps { `addn` `subn` `muln` `divn` } and { `addi` `subi` `muli` `divi` }.

There is a third category, which need encompass just one instruction.

A sequence alone offers insufficient structure with which to express every program. For example, it is both inefficient and inadequate to unravel repetition into a long sequence. For one thing, it makes the program longer — often much longer. Second, it is frequently not possible to know in advance when the end will come. A search can be reduced to repeated inspection, but it ends when the quarry is found. It is impossible to know how many iterations will be needed *when writing the program*.

We need an instruction which selects whether to 'branch' back to an earlier point in the program or to proceed. The decision can be inferred by the state of the register at TOS: zero or not. A second operand will be a signed number of steps to go back (or forward). For example, to repeat a sequence of instructions, perhaps to sum an array, we first load the TOS with its length. After adding in the latest contribution, we simply decrement the TOS then branch back, *unless* we have reached zero.

The two-operand instruction we need is called a "conditional branch", oft-abbreviated to `cb`.

Because `cb` can refer to the stack for its decision, only one operand need be explicit.

To summarise, a typical stack-architecture instruction set comprises:

— { `push pop` }	one-operand	destination/source
— { `cb` }	one-operand	signed offset
— { `addn subn muln divn` }	zero-operand	
— { `addi subi muli divi` }	zero-operand	
— ...		

Actual instructions that a processor can execute are encoded in binary, and are thus somewhat less than transparent. As a result, *mnemonics* are written instead for human consumption, such as `addn` for "add two natural numbers together". Here is a program 'snippet', which performs $z \leftarrow x + y$:[20]

```
push x
push y
addn
pop z
```

x, y and z must all be natural numbers and "bound to" (allocated) a register in main memory.

Repetition can be expressed easily by attaching a label to the first instruction:

```
start:      push count
            ...
            subn #1
            cb   start
            ...
```

`count` must refer to a variable holding the number of iterations required, denoted by a natural number. A '#' denotes a value (literal number), rather than a variable.

20 '$\leftarrow$' is read 'becomes' and is commonly used to denote assignment. '=' would be wrong as it merely asserts equality, rather than an operation. An arrow is appropriate because it indicates the direction of data flow.

Expressing a program in this way is said to employ "assembly language" and requires translation into the (binary) numerical version executable by a machine. Each mnemonic must be replaced by an opcode and each variable by its bound address, each offset calculated. The task is performed by a dedicated program called an 'assembler', or by a 'compiler', which can also translate a 'high-level' language, like *C* or *Java*.

If there are just fifteen possible operations, as will be the case if we support only natural and integer numeric data types, the opcode requires just four bits — half a byte, or one 'nybble'.[21] Most machines cannot address any element of memory smaller than a single byte. While we could cram two instructions into each byte, we typically prefer to house just one, leaving a nybble free.

Operators (like `addn`), which normally locate both their operands at the TOS, can instead refer to the otherwise vacant nybble next door, allowing simple and efficient implementation of decrement and increment by a small signed value. Zero might indicate reference to the stack instead, since it would not normally constitute a useful, or even (with `div`) legitimate, operand.

`cb`, which can refer to the TOS for its decision, branching when it is *not* zero, can often make use of a signed offset small enough to fit within a nybble.

We have seen how each variable is mapped to an address within data memory. A value, which remains constant while the program executes, must somehow also be defined. Some method is needed by which any value, of any type, can be created within the code. We've already seen how a small value can be accommodated within the instruction. To create a larger one, we need another instruction.

If we can shift the binary equivalent of a single hexadecimal digit into a dedicated register then we can assemble any desired value, signed or otherwise. The assembler can generate the desired sequence, so we need only give the complete (signed) version in our assembly language program. Hence:

```
addi #-2
```

will, using two's-complement representation, be translated into:

```
shl  #F
shl  #F
shl  #F
addi #E
```

This might seem inefficient, but in fact code rarely requires large or negative constant values, so `shl` instructions in fact remain few in number, adding little to the size of any program.

---

21 Many machines also provide arithmetic instructions for 'floating-point' numbers as well. Recall, these are a very poor substitute for the 'real' numbers encountered in science and mathematics. The problem arises from the infinite 'real' domain between any two values. A finite machine can only offer a finite domain, meaning any 'real' value is only approximated. So-called "rounding errors" can sometimes grow to such an extent that a result can be, in effect, random.

Now that we have 32- and 64-bit machines, we usually have the option to remain safe and employ only integers.

Fetch-execute cycle

The true magic of a computer is in its ability to execute a program. That program is stored as a sequence of instructions in memory, regardless of its structure, which emerges according to the use of cb.

When a typical computer is "powered up" (switched on), whatever occupies the register whose address is zero in program memory is loaded into the processor, which will attempt to execute it. Each subsequent instruction is then fetched and executed. This is the essence of what a computer is built to do, and is called the "fetch-execute cycle". It will carry on until "powered down" (turned off).

To facilitate the fetch-execute cycle, the machine employs a dedicated register, usually called the "program counter" (PC), which is cleared upon 'power-up' and incremented after fetching each instruction. It always holds the address of the next instruction in program memory.

A simple calculation suggests that no program can possibly take more than about a second to execute. Suppose you have a gigabyte (billion bytes) of program memory and a processor which operates at a gigaHertz (a billion cycles per second) and executes one instruction per cycle, on average, and is thus fairly typical for a personal computer at the time of writing. You might well think that, as a result, even a program which completely filled program memory would terminate in one second.

Yet many an internet server continues to run all year round.

The answer to this apparent paradox is that programs have structure, including much repetition. They have loops within loops, as well as much selective behaviour.

But every computer tirelessly implements a fetch/execute cycle for as long as it remains powered, which, given the fabulous reliability of electronic technology, may be a very long time indeed.

Processor organization

Instruction execution pipeline

It has long been known that the most efficient way to manufacture anything is via a "production line". While it may take many hours of labour to assemble a car, one may roll of the end of a modern production line every few minutes. The trick is to divide the process into p smaller ones and run them *in parallel*. The output of one becomes the input of the next, dividing unit production time by p.

The process of executing an instruction can also benefit from the use of a 'pipeline'.

Figure 4.49 shows one way to subdivide execution into five 'stages'.

Some parallelism is exposed even within a stage: in Stage Two, during the second clock cycle, the operation can be decoded (from the 'opcode') while two registers at the TOS are read. In the fourth cycle, however, *either* data memory is accessed *or* a branch target address is written to the PC.

We never need to do both at the same time.

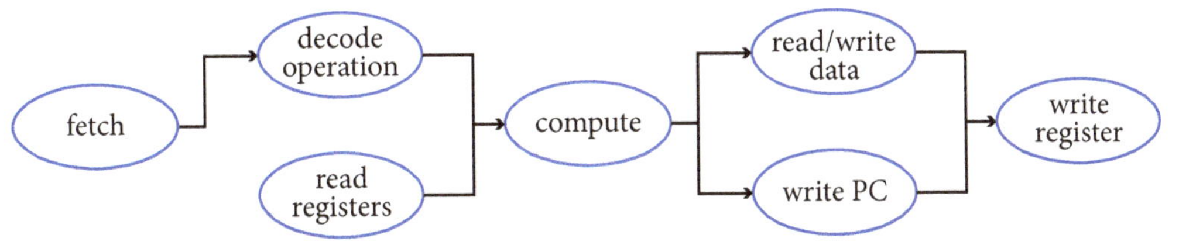

Figure 4.49 A possible five-stage pipeline for instruction execution.

Decoding the operation means deriving a collection of control signals that take effect in later stages. What happens in Stages Three, Four and Five thus depends on the nature of the operation. For example, the fifth stage will only become active when executing an arithmetic or `push` instruction.

Note that we *always* read the two registers at the TOS. If their content is not required, it is simply discarded. Nothing is lost or changed. When an arithmetic instruction is decoded, on the other hand, Stage Three uses their content as input for the operation signalled. The result is fed through Stage Four and written into the register now at TOS in the fifth cycle. The inactivity of Stage Four in the fourth cycle is of no concern at all. All we care about is the *average number of clock cycles per instruction* (CPI). This will be close to one, even though each instruction takes several cycles.

The only way an instruction might not complete execution in a given cycle is when there is competition for the same resource in some stage. The only resources shared between two stages are the registers. Suppose we have two arithmetic instructions in sequence: the second cannot be allowed to enter Stage Two until the fifth cycle of the first, because the TOS will only then contain its result.

Since arithmetic operations often follow in sequence, a three-cycle delay is unacceptable. However, a one-cycle delay is inevitable and intrinsic to computation. Stage Three can in fact deposit its result directly in the current TOS, so that only a single cycle delay is necessary. Stage Four can go.

Remember that each stage is busy (almost) every cycle. At any one time, each stage is "working on" a different instruction. It is useful to draw a "time/space diagram" where time progresses from left to right, while the instruction concerned varies downward. For example:

cycle #	1	2	3	4	5	6	7
`addi`	fetch	read	compute				
`muli`		fetch	o	read	compute		
`divi`			fetch	o	o	read	compute
...				fetch	o	o	o

One 'bubble' is generated on each arithmetic instruction, increasing our CPI.

There seems to be a conflict between stack architecture and pipeline organization. While a stack architecture is efficient at the instruction level, it appears to lead to inefficient use of a pipeline.

Instruction fetch

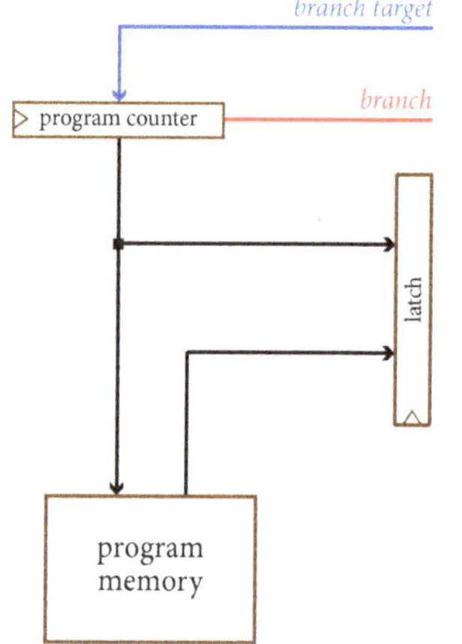

Figure 4.50 Pipeline Stage One: instruction fetch.

At the heart of any contemporary computer is the Program Counter (PC), which carries the address, in dedicated program memory, of the next instruction to be executed. It is designed so that its content is 'cleared' — *i.e.* reset to zero — when the machine is powered up. This has the effect of causing execution to begin at the base of program memory (the location whose address is zero).

Upon each clock tick, the PC increments, to identify the next instruction. It may instead load a new value, identifying a "branch target", according to a signal.

The program memory must return the next instruction within half a clock cycle, so that it may be latched ready for the next stage to decode on the clock tock.

It's actually more likely that the device employed for program memory will require two clock cycles to return the content of a register. In that case, clock ticks must alternate between the PC and latch. These are equivalent to the tick and tock of a clock signal of exactly half the original frequency.

Decode, read registers and compute

Figure 4.51 shows how arithmetic instructions may be carried out by Stages Two and Three of a pipeline. In Stage Two, the instruction is broken into opcode and 'immediate' operand. The opcode is decoded, via a dedicated logic unit, into a number of control signals (depicted in red), most of which are fed forward. The operand nybble is copied into a "shift register", whose contents shift left four bits each cycle. A single control signal might cause it to clear (to zero) rather than shift. Any larger operand is built up via `shl` instructions which each contribute another nybble.

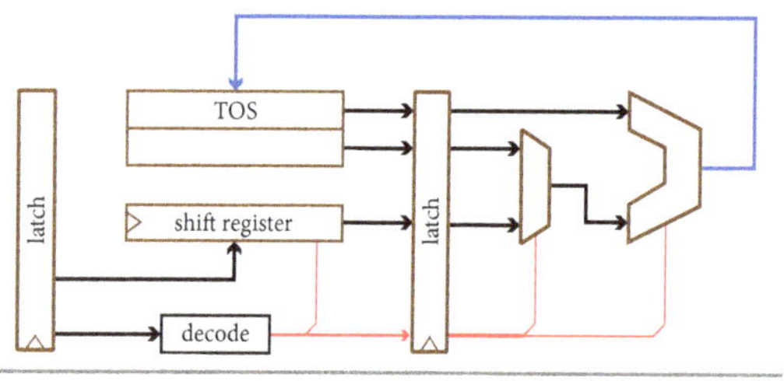

Figure 4.51 Arithmetic computation in a pipeline.

This mechanism allows an 'immediate' (constant) operand to replace one on the stack. A decoded control signal might switch one input of the ALU in Stage Three between the two.

Whether both registers at the top of the stack or just one and the shift register are carried forward to Stage Three ('compute'), we must make sure the TOS is not read again until it's been written. This might mean stalling Stage One.

A bubble is then introduced in the time/space diagram.

To access the tos, all we need do is activate the register which *currently* holds that privilege. A counter (not shown), retains the correct index into the processor register file. This is decremented (by one or two) each time the stack is read and incremented when it is written. When it becomes zero, the topmost items are copied in from data memory filling the register file, as far as possible. When the register file is full and something is pushed on, its contents are flushed into data memory. Given a file of n registers, the topmost n items on the stack are said to be 'cached' inside the processor.

Now we can return to the apparent conflict between a stack architecture and pipeline organization. This emerges simply because it is highly likely that a series of arithmetic operations will be necessary in order to evaluate an expression. One will follow another, with the result of the first frequently forming one operand of the next. It is pretty clear that there's nothing wrong with using a pipeline, we just need to design it with respect for the fact that one instruction will often need to use the result of the last.

If we recall that each access to program memory is likely to take two cycles, because of the need to decode an address (among other things) we can see that the problem disappears if both Stages Two and Three each take one cycle. All that is necessary is the correct timing, which is easily arranged.

We do, however, obtain less parallelism, and thus a higher cpi — *i.e.* a lower performance.

Branching

To complete a branch to either an earlier or a later instruction, we need to add a signed offset to the program counter. Since we have an alu already, in Stage Three, we merely forward its content up the pipeline, *after* incrementing it. (The offset is from the location of the instruction that would be executed next, if the branch is not taken.) The offset itself is found in the spare nybble within the instruction or, if it cannot fit, in the shift register, formed via one or more prior `shl` instructions.

The `cb` instruction requires two things fed back: the *branch* signal, causing the pc to load new content, and the *branch target*, which forms that content. The simplest decision that we might employ to decide the branch is whether the tos is zero; for example, when a count down has ended. This is easily accomplished, using a single multiple-input nor gate. The target is the sum of program count and offset.

Given a two-cycle Stage One, there should be no delay in instruction execution, whether the branch is taken or not.

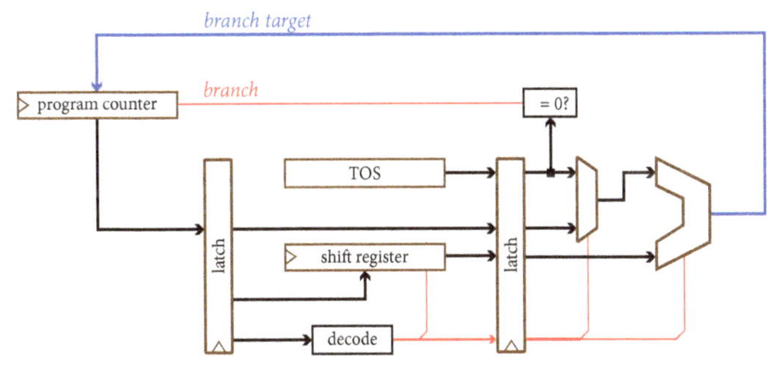

Figure 4.52 Branch computation in a pipeline.

Push and pop

A 'load/store' architecture implies that only one instruction loads something from data memory and only one stores it there. To evaluate an expression, we first load every variable concerned, perform the necessary computation and then store the result afterwards. The 'load' instruction for a stack architecture is `push` and the 'store' one `pop`.

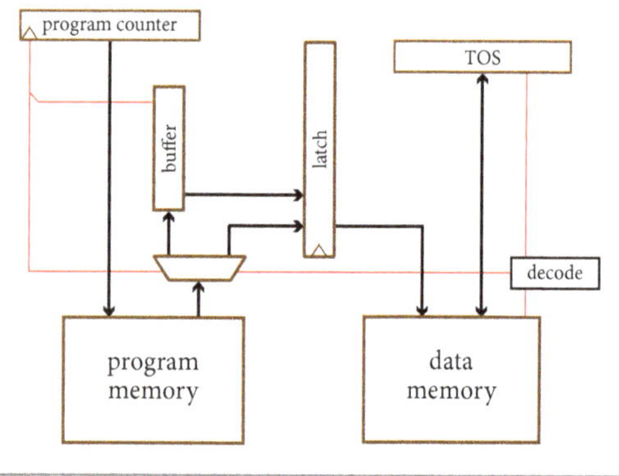

Figure 4.53 Possible support for `push` and `pop`.

Every `push` or `pop` instruction will be followed by the address where the variable concerned resides in data memory. That address will be stored in program memory, immediately after the instruction itself. We need to add a switch to Stage One, so that what is returned from program memory may be passed forward to Stage Two appropriately.

It's easier to always load an item of the same size. Since an instruction is typically much smaller than an address, we might load a bunch of them in one go. Here, we've suggested using a 'bytecode', where each instruction fits within a single byte. Currently, an address often comprises eight bytes (sixty-four bits). Hence we either load one address or *eight* instructions.

Given that only one instruction is sent forward each cycle, we need somewhere to keep a complete bunch of instructions. They might form a queue in a dedicated buffer, as shown in Figure 4.53, and be forwarded one per cycle. Since we now only need to load instructions when the buffer is empty, we need to control the advance of the Program Counter (PC).

The PC is now advanced only when the instruction buffer is filled or an address read.

To summarise, when a `push` or `pop` is passed forward to Stage Two, it is decoded, causing control signals to 1) switch program memory output onto the data memory address path, 2) increment the PC, 3) tell the TOS register and data memory the direction of data transfer (on the subsequent cycle).

Since instructions need only be read every eight pipeline cycles, `push` and `pop` typically require just two: one to decode plus one to transfer. The address will usually be ready and waiting.

The reader must understand that the organization presented above is subject to many assumptions, and a limited set of aspirations. Real computer organization will tend toward greater complication.

It serves to illustrate methodology only.

Memory organization

Memory requirements

Perhaps the most obvious requirement we have of machine memory is the same priority we have for our own. We don't want to forget things, which is perhaps why photos and movies are so important to us. To use the technical term, we want our machine memory to be *non-volatile.*

We've thus far discussed two different kinds of memory: static and dynamic. The latter required continual refreshing; otherwise, the information contained in the register would flow away — it was otherwise volatile. Static memory — registers composed of flip-flops — is sadly also volatile; if the power supply fails, the content will, again, disappear down the drain. We need something else.

Memory must also be *portable.* These days, it is common for professional people to work not just in the office, but at home and even on the move. Information must be accessible everywhere, to the extent that we should ideally no longer care where it resides. Older books speak of portable media as a solution. These still exist but are less convenient, though we still see the occasional stick or card.

Their small size has become more of a disadvantage; they are easily lost.

The solution begs a great deal of trust in automation. Our phones, pads and laptops automatically move our files to a server via the internet, even as they are created, making them available with any device we possess and to anyone with whom we choose to share them. Information has never been so portable and probably can never become more so. But this threatens another requirement.

Memory must be *secure*, against threats both natural and human.

Choose your medium — perhaps a blu-ray disc, for your wedding video — then pop it in a drawer for twenty years. You might be lucky; it might play fine; but then again... Material degrades over time; accidents happen; and technology changes. (Will blu-ray players still exist, later in life?)

Like every other tool we humans have ever invented, memory requires maintenance.

You will often (hopefully) hear people speak of the need to make a 'backup' copy of some digital recording. But it is not that simple. Much of our information is highly dynamic — it changes rapidly with time. What we need is automated *archival*, allowing us to access 'snapshots' over time and thus to easily recover a desired version of some or other file. The trouble is that our hardware is so reliable, and thus information loss so rare, that we tend to overlook this need, until it's too late.

Given that any system can be reduced to a collection of just files and processes, it's well to point out that processes too are worthy of archival. As with files, it can be vital to restore the state of any process to one it previously possessed. To this end, 'checkpoints' can be identified, and recorded.

The last two requirements of memory to be met are the greatest *speed* and lowest *cost.*

There's a trick in reconciling these two, which naturally conflict.

Memory hierarchy

Recall that computation comprises three functions: processing, memory and communication (i/o). While one day we may realise all three in a single system, a contemporary computer is typically made up of three separate subsystems, each performing a single function. An inherent limitation of any such machine is imposed by the *internal* communication thus required between processor and memory.

(Recall that external communication is achieved via "memory-mapped i/o".)

In separating processing and memory, we have created an additional problem arising not from any computational task at hand but from the way in which we intend to perform it. The speed with which can now compute will be limited by the speed with which we can access memory. But the degree to which we can exploit the automation of computation will certainly depend upon the cost.

It should come as no surprise that the fastest memory devices are also the most expensive. So, we have a problem in reconciling our need for speed with our equally important need for low cost.

Salvation can be found in taking greater care in defining *precisely* what we require. We need:

— the lowest delay *per reference*

— the lowest cost *per bit*.

To deliver both the greatest speed and the lowest cost *in the same device* is quite a challenge. Fortunately, there is a way to achieve both aims, simultaneously, by combining *different* devices. We employ a *hierarchy*, where both speed and cost decrease as we move downward, while size increases (Figure 4.54). For a given item (address), we look first in the topmost device. If we fail to find it, we refer the request downward. This continues until we reach the lowest device, where the item *must* be found.

Recursion must always terminate at some point, in the real world.

All we have to do is keep the items we are most likely to need next 'cached' at the top of the hierarchy. Those of somewhat lower priority may be kept at the second level, and so on. Everything else resides in the basement. But how do we know just what we are likely to need next?

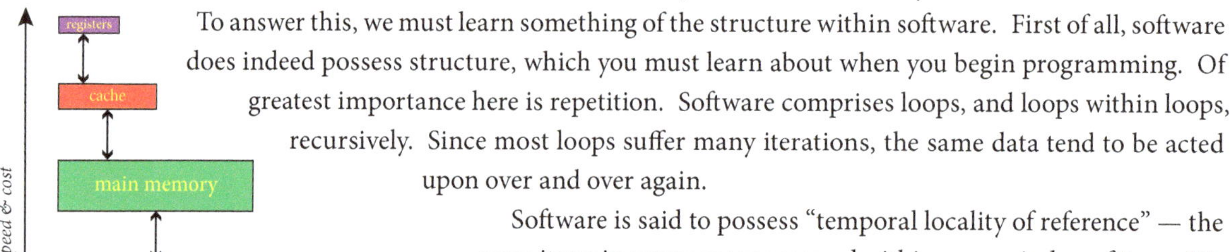

Figure 4.54 A memory hierarchy.

To answer this, we must learn something of the structure within software. First of all, software does indeed possess structure, which you must learn about when you begin programming. Of greatest importance here is repetition. Software comprises loops, and loops within loops, recursively. Since most loops suffer many iterations, the same data tend to be acted upon over and over again.

Software is said to possess "temporal locality of reference" — the same items in memory are accessed within some window of time. We can thus create *spatial* locality by caching those items together.

Caching allows us to exploit the technology that permits the lowest cost in one device and the greatest speed in another to achieve both the lowest cost per bit and the lowest delay per reference.

If the software we run exhibited no temporal locality then caching would not work.

All we need do is engage a caching strategy that reflects the fact that the same things tend to be needed repeatedly. One such strategy is to allow the latest item read from below to overwrite the item *least-recently used.* (It will only be recovered from below if not already present.)

It's easy to show that the cost can be kept near to the minimum as long as the size of the device scales sufficiently greatly as we head for the basement. If the cost per bit at level i is c_i pence, and the size is s_i bits, then the total cost C for all n levels is:

$$C = \sum_{i=1}^{n} c_i . s_i$$

The mean cost per bit $\bar{c}$ is thus:

$$\bar{c} = \frac{C}{S} = \frac{\sum_{i=1}^{n} c_i . s_i}{\sum_{i=1}^{n} s_i}$$

If the size s_i is scaled inversely with cost c_i, we will always end up with $\bar{c} \approx n \times c_n$, where n indexes the cheapest (basement) level. As long as the scale factor between levels is much greater than n, this is fine. For example, four levels with a scale factor of one hundred will result in a mean cost per bit that is approximately four times that of the cheapest device. Each bit in the most expensive (topmost) device may cost a *million* times more than one in the cheapest, so we are actually doing very well.

As a rule of thumb, we might spend about the same on each device in the hierarchy.

To calculate the mean delay $\bar{t}$ per reference, we need to know the "hit rate" h at each level:

$$\bar{t} = \bar{t_1} = h_1 . t_1 + (1 - h_1) . \bar{t_2}$$

This is called a "recurrence relation". To calculate $\bar{t} = \bar{t}_1$, we must know $\bar{t}_2$; to calculate that, we must determine $\bar{t}_3$; and so on. Recurrence stops at the basement level, where $h = 1$.

The first term — $h_1 . t_1$ — describes the contribution from those references that score a 'hit'; the second one describes that from the remainder. Suppose we take a sequence of one hundred references, where the top hit rate is 90% ($h_1 = 0.9$) and the delay (t_1) 10ns. For ninety references, we shall wait a total of 900ns. If the mean delay for the remaining ten, which must be referred onward and downward, is 100ns then their contribution will be 1,000ns, meaning that the mean overall delay ($\bar{t} = \bar{t}_1$) is 19ns.

We find that, in order to obtain $\bar{t} \approx t_1$, we need a high hit rate of 95 to 99%. This can be achieved, but it does depend on the structure within the running program.

In the absence of appropriate temporal locality, performance will decline dramatically.

Memory-mapped i/o

Responding to events

Beyond the normal flow of control, where it obeys instructions in a 'main' program in the sequence in which they're found, a processor must also be capable of *reactive* behaviour — it must respond to events. Some of these events will occur within the computer system, and some without. For example, a painfully slow mass storage device might just have located and despatched some requested data, or a key has been pressed on an external keyboard.

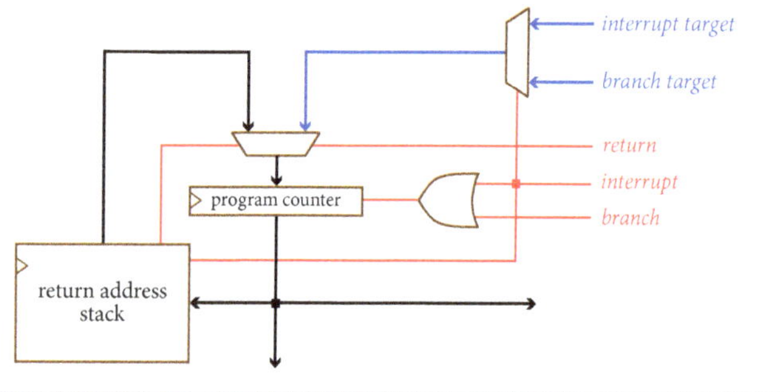

Figure 4.55 Interruption of normal control flow and return.

Any such event can be signalled by asserting a logic '1' on a wire, in time for the next clock tock, whereupon normal control flow is interrupted. The start address in program memory of an "interrupt service routine" (ISR) must replace the current PC content, which is pushed onto a stack. The final instruction of any ISR must be "return from interrupt" (`rti`), which pops an address back.

Return addresses must be stacked because any ISR can itself be interrupted and we need to nest event response as a result. For the sake of speed, the stack may well be implemented using a dedicated register file, indexed (using yet another register) to allow identification of the current top (TOS).

Figure 4.55 depicts our revised instruction fetch mechanism (Stage 1 of our pipeline). The 'interrupt' signal originates beyond the processor with a device of some kind. Both 'branch' and 'return' signals result from decoding an instruction — `cb` or `rti`, respectively. (Note that there is never a return from a branch.) The OR gate ensures that the PC loads on the next clock tock whenever either branch or interruption occurs, while the two switches are set to correctly identify the new content. The 'interrupt' signal causes a push onto the stack, and 'return' a pop from it.

One thing we must take care about is that the PC is *always* incremented, even just before it is copied onto the return address stack. When that address is popped back, it must identify the same instruction it would have done before: the one *after* the last instruction executed in the interrupted routine.

Each interrupt target (start address of an ISR) is selected from an "interrupt despatch table" (IDT) using an index called the "interrupt vector" supplied by the device requesting interruption. The IDT, along with the ISRs it identifies, forms the greater part of the kernel of the computer's "operating system". Together, they make it possible for the various devices which make up a system to communicate.

They also form its greatest vulnerability. Should an invasive program run, it might leave its own code behind, intercepting interrupt service. The system is then said to be 'infected'.

Keypad

To show how we might connect external devices to our computer, we shall first consider a simple keypad. Suppose this has sixty-four keys, which we might imagine laid out as an eight by eight array.

If you break up an old keyboard, you'll find a plastic membrane inside separating one bunch of conductors running below another. We can imagine the two aligned at right angles so that each key lies above an intersection between one vertical wire and one horizontal. Some form of spring keeps it raised. When we apply pressure, it will push the upper conductor down onto the lower one, making a connection. Any potential difference between the two wires will cause a current to flow.

A single byte is more than enough to record a choice from sixty-four keys, but is the least normally addressable in memory. We shall divide the six bits required into two 'fields' of three: one to record the row, the other the column of any key pressed. One of these — let's say the row field — forms a counter, continually counting modulo eight. This is connected to a binary decoder, which asserts the output (of eight) corresponding to the count. Rows are thus scanned, with a logic '1' periodically on each. Should a key be pressed, that '1' will also then appear on a column. We encode which in binary and latch it in the remaining field.

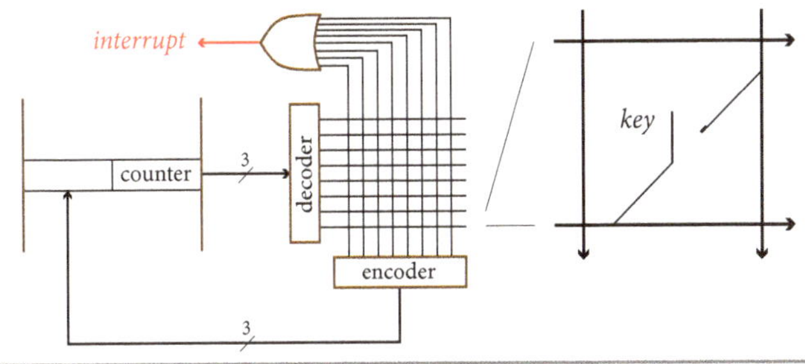

Figure 4.56 Memory-mapped keypad.

Together, the fields record the key pressed.

It remains only to interrupt the processor so that something can be done in response. A suitable signal is easily derived via an OR gate, as shown in Figure 4.56.

There is no need to worry that the count might have changed by the time the interrupt service routine (ISR) reads it. Mechanical processes are a million times slower than electronic ones, so the count can increment on a timescale far beyond the 'latency' and that required to execute a few instructions.

A keypad ISR must accomplish two tasks. It must first map the key code recorded in the memory-mapped input register to the correct *character* code, perhaps using the 7-bit American Standard Code for Information Interchange (ASCII), or the 8-, 16- or 32-bit Universal Translation Format (UTF) for the much newer Unicode Character Set (UCS). Either way, whoever composes the routine will also require knowledge of the keyboard arrangement. Second, the character must be enqueued within a keyboard 'buffer', to await consumption by some process making use of this particular device.

The ISR plus a routine that initialises the keyboard buffer form a fairly typical "device driver".

Display

Any imaging display connected to the computer will require enough dedicated "video memory" to represent one 'frame'. This memory is special in that it can be accessed by an external device as well as by the processor. Hence, it is often referred to as 'dual-port' memory. The display controller is essentially a counter which cyclically addresses each word in video memory, which records the brightness and colour of a 'pixel' within the display.

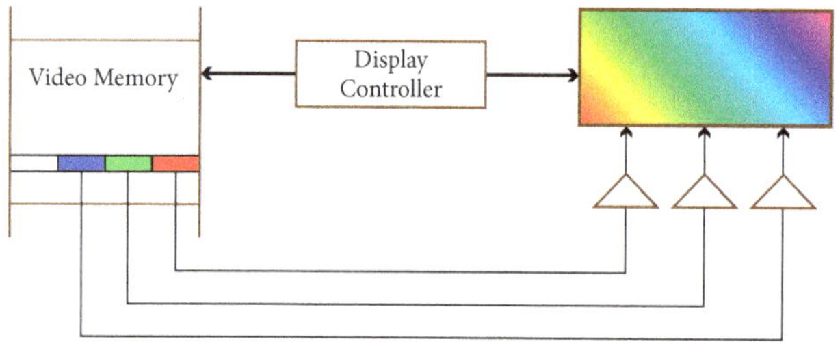

Video Memory
Display Controller

Figure 4.57 Memory-mapped display.

Each pixel record typically comprises three fields, of equal size: red, green and blue. The number of bits employed will decide the resolution possible, in both brightness and colour. Whatever remains available may be used to label the pixel.

Each field will require digital-to-analogue conversion for the display to operate.

Resolution must be defined in three separate dimensions: brightness/colour 'depth', spatial (the number of pixels horizontally and vertically) and temporal. The latter is defined by the frequency with which frames are read and converted. The human eye processes discrete frames at a rate of around ten to fifteen per second. To avoid judder (and headaches), the display frame rate should exceed twice that. For decades, movies were captured at 24 fps, but a new standard warrants 60 fps. Digital projection is now also improving both depth and spatial resolution, perhaps beyond perception.

Application

4.12 Given the following information about a three-level memory hierarchy, calculate both $\bar{c}$ and $\bar{t}$:

LEVEL	c p/MB	s MB	t ns	h
1	512	16	1	0.95
2	16	512	10	0.9999
3	1/16	65,536	10^7	1

Explain in detail what determines the final column, and how.

4.13 Try drawing Figure 4.54 to scale, when each level is a hundred times the size of the one above.

If each level offers a delay a hundred times that by the level above, draw a comparison with ordinary life before electronic communication, when the fastest message was a paper memorandum from a colleague elsewhere in your workplace, delivered by errand boy.

4.14 A computer display has an aspect ratio of 16:9. Assuming 8 MB[22] of dual-port video memory is fitted, what is the highest spatial resolution possible with 24-bit colour depth?

A computer has memory that is 64 bits wide, which can be read in 3 clock cycles, and runs at 1 GHz. Its processor has 8-bit instructions, and executes one every clock cycle. A camera has direct memory access and the same spatial resolution and pixel depth as the display.

At what frame rate could the camera update the display without significantly affecting program execution? (Assume only the camera and processor share access to memory.)

Televisions and computer displays are becoming indistinguishable. (Most televisions now have both a camera and internet access.) Whether the user watches conventional broadcast content, via an aerial, or programmes 'streamed' over the internet, the TV must possess a "frame buffer" (video memory capable of holding at least one frame).

Comment on the demands on memory performance in a "4K UHD" TV, where spatial resolution is 3,840 by 2,160, pixel depth is either 30- or 36-bit and the display is redrawn at 60 fps.

When connecting a 4K UHD Blu-ray player to such a TV, what capacity (in bits per second) must the cable offer? Compare that with typical internet download performance and comment.

4.4 The future of computing

Parallel computing

For the first fifty years or so of our building computers, faster processing meant improving the one and only processor. The principal means by which this was accomplished was to reduce the size of the fundamental component — the humble switch. On a macroscopic scale (when you could see the thing), a smaller switch did indeed mean a faster one. It continued to work as the scale dropped towards the microscopic, and life was rosy... but then it stopped, at around a micrometre ('micron' or μm).

22 In the computer world, 'gigabyte' (GB), 'megabyte' (MB) and 'kilobyte' (KB) traditionally refer to the nearest power of two; hence, for example, '4 KB' would mean 4,096 bytes.

One reason why things began to change is that a narrower conductor actually has greater resistance, not less. Any conductor also forms an aerial, which both transmits and absorbs electromagnetic energy, so interference and noise become real difficulties. Greater and tighter integration of devices becomes vital if relative distances are not to grow, making resistance an even bigger problem. Lastly, as clock rates move into GHz territory, even the speed of light can impose limits.

For a while lengthening pipelines, and even employing more than one, seemed to offer a way of keeping performance racing ahead. But longer pipelines face diminishing returns.

Ultimately, there is only one way of increasing performance almost indefinitely. Instead of having but one processor, we employ *n* of them, which means they must communicate if they are to co-operate on a common task. If we can achieve *scalability* then performance will grow linearly with *n*.

A task may demand either more processing (of more data), or faster processing. We thus need performance, in both regards, to scale with it. It is clearly not just the architecture of the hardware whose performance must scale with the task. The software architecture must do so as well.

Given that any running program forms a 'process', it is only natural to decompose it into a number of procedures, each of which forms a subsidiary process when it executes. Any programmer, trained at any time during the brief history of the discipline, will be familiar with the recursive composition of procedures and their *sequential* execution. They may also compose and execute concurrently.

Sequentially executing processes communicate via (sequential) access to common memory (variables). Processes which execute concurrently require synchronization to communicate securely, and must be provided with dedicated channels, mapped onto hardware 'links'.

If we construct our software from many communicating processes, we can map them onto a smaller number of communicating processors. We require *Communicating Process Architecture* (CPA) in both. Note that CPA mandates excess concurrency in software over hardware.

Fortunately, one of the greatest contributors to Computer Science, Professor Sir Tony Hoare, has provided a sound formal basis on which to build, with the theory of Communicating Sequential Processes (CSP). When you set out to build a house, it is well to first establish a secure foundation.

Since its beginnings during World War Two, the implementation of both hardware and (particularly) software has raced ahead, to meet exponentially rising demand. While hardware technology has led the way, as a result of revolutionary progress in understanding the physics of the solid state, software technology has lagged behind, for two principal reasons: the enormous complexity of any program (or process) of significant size, and the simple fact that there are many possible approaches, each with their vociferous advocates. The traditional scientific approach has not lent itself well to the development of a proper understanding of software, and thus of dependable technology. Programming has also lacked a secure foundation, with adequate education and training to underpin a formal discipline. Many are the advocates of solutions to both problems. Perhaps some, even many, may be right, but who can tell the voice of reason in a crowded room when all are shouting.

In such circumstances, we can only turn to established leaders, like Hoare.

In the UK, during the late 1980s, a combined hardware/software solution was developed based on Hoare's CSP. The 'Transputer' and its 'occam' programming language was convincing but failed commercially for two probable reasons: the world wasn't ready for something so radically different, in so many ways; neither did the world yet widely need so much power, so scalable and so affordable.

The world is certainly ready for scalable high-performance computation now that both products and production are becoming fully automated. The program controlling a mobile phone passed a million lines of code by *c.* 2010, and already ran on multiple communicating processors. A typical car now sports dozens of computers. Many such 'embedded' applications accomplish formidable tasks.

Thirty years ago, only spooks, weather forecasters and a few research groups needed it.

A transputer (the term is now generic) is a complete computer, incorporating a means of synchronous communication with a number of others, allowing them to be interconnected to form a 'surface'. Network topology depends upon the number of links possessed by each node, but is usually symmetrical and unbounded, yet finite, such that the 'diameter' scales slowly with size.

A simple example is that of a 'torus' (like a bagel) as shown in Figure 4.58.

While the notion of a computing surface might suggest something large and power hungry, it lends itself well to large-scale integration. A thousand-node surface might be possible already on a single chip.

Replication simplifies both hardware and software design.

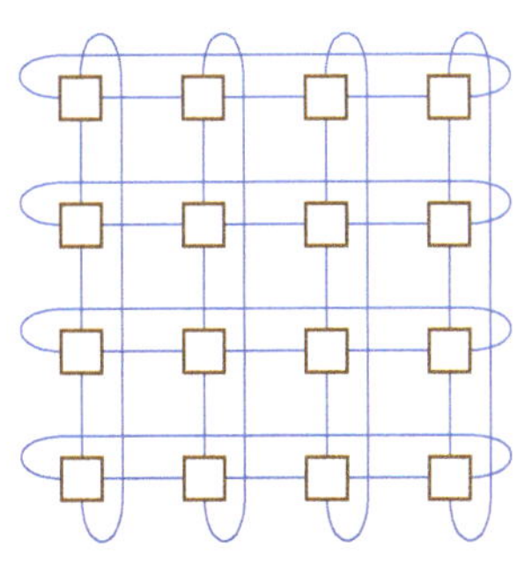

Figure 4.58 A torus computing surface.

One underlying characteristic underlaid the technical success of transputer networks: as a computational task scales, its internal 'compute/communicate' ratio tends to remain the same. This allows the same device to serve in networks both large and small. One principal difficulty contributed to its commercial failure: it required programming concurrency, which appears to involve little that is unfamiliar, while hiding the potential for entirely new and catastrophic errors.

The problem of populating many processors with many more processes is not difficult. While composing a distinct procedure for each process would clearly scale the workload, along with the task, replication of a single program, subject to parameters can actually reduce it. What is required is a suitable concurrent algorithm, where conventional expertise (and textbooks) offer only the sequential.

Perhaps the most insidious error newly invited in with concurrency is 'deadlock'. This has been known and discussed for decades, since concurrency has in fact always been present. Even in the simplest of systems, there is more than one agent at play, even if they're only the user and their computer. The problem was merely hidden within the operating system, provided by a special category of 'systems' programmer, who was expected to understand both the hardware itself and concurrency.

In fact, we all live with concurrency all the time and form a communicating process ourself. So it is hardly surprising that we come face-to-face with deadlock now and then. Remember what happens when you attempt to pass through a doorway when someone is trying to come the other way? You either follow the "after you" or, if you're a bully, the "after me" protocol. Neither works.

Even after several invitations by either party, a collision is distinctly possible.

A variety of protocols exist which exclude deadlock, backed up with formal proof. A simple one requires a regular array of processes which each perform all their communication in a constant cycle. Such Cyclic-Ordered Processes were proposed by another giant among the founders of Computer Science, Edsger Dijkstra, and his colleague, Carel Scholten, long before they were really needed. More generally applicable is the 'master/servant' protocol of yet another great, Per Brinch Hansen, now more commonly known as 'client/server', and widely used already across distributed systems.

There is no space here to properly describe such protocols. We merely note that formal methods exist to allay any fear of concurrent programming. Nor should there be any fear of tackling the mathematics they require. An engineer building a bridge does not usually need to repeat the formal analysis made by his ancestors. Instead, all he need do is to apply *design rules*, which come already proven.

A software engineer can do the same.

Whenever a programmer uses a high-level language — one in which rules of syntax apply — they are already, in effect, applying design rules, which exclude many errors in exchange for the denial of some legitimate (but generally less transparent) programs.

Ideally, all design rules would be built into the programming language, making it possible to compose only those programs which are 'correct'. This remains a matter for research, but that should also take account of whether or not automation should extend even unto programming. It has long been argued that the ultimate programming language is not for the implementation of software, nor even for its design, but instead only for its *specification*, which might automatically generate both.

We are only a half century or so into information technology.

It took three times that long before its mechanical counterpart became simple and safe.

While it may take some time to mature, all the necessary knowledge exists to make scalable parallel processing a reality, enabling unbounded computational resource, with all that it implies.

Unbounded automation

There was a time when experts genuinely believed that the world really didn't need computers, despite the fact that they had been involved in their development for the purpose of defeating Fascism. In the wake of the ultra-secret revolutionary accomplishments at Bletchley Park, public development continued at three centres in the UK: Cambridge, the National Physical Laboratory (NPL) and Manchester University, where the world's first general purpose, all electronic, digital, stored-program computer ran in June 1948. Professor Douglas Hartree, in charge at Cambridge was of the opinion that the three computers under construction would be all that would ever be needed, or afforded. Charles Darwin (grandson of the author of *The Origin of Species*), the head of NPL, believed that a single machine would suffice. Across the pond, the master of computer construction at Harvard University, Howard Aiken, thought that the US might need a few more: a half dozen perhaps, "hidden away in research labs".

They thought of the computer as something which merely acted on numbers. They remained in ignorance of the visionary understanding of Ada Lovelace, almost a century earlier. She saw how, since numbers can encode anything, a computer could process anything, including all that we perceive.

No-one foresaw the potential; arguably, it is still grossly under-rated.

The most obvious benefit of the availability of cheap, compact, powerful processing is the ability to automate any repetitive task, manual or mental. Only the most profound human skill is excepted, such as medical diagnosis and driving a car (safely). Many things which humans find difficult turn out to be relatively easy for a machine, such as flying a plane or complex calculation.

Of course, you might not perceive it as a benefit if you lose your job. Mass unemployment was one result of the first industrial revolution, which automated just one phase of the production cycle (assembly). We will soon be able to automate *all* phases. "Lights-out" factories are a reality already. Another difference in the unfolding second industrial revolution is that mass production will no longer be necessary in order to take advantage of the technology. "3-D printers" are but the start.

In the last section, we argued that computers may soon be able to both design and implement their own software, leading to the need only to specify (clearly and unambiguously) what it is you want to achieve. Their system software may be capable of learning your body language so that they know when they are being addressed. Whether we speak politely to our new morally-acceptable slaves will be a matter of taste. They will be incapable of taking offense.

Note that it is typically neither necessary nor desirable that our automata take the form of "walkie-talkie" humans or that they should possess consciousness. The latter would defeat the whole purpose since their slavery would cease to be morally acceptable. They would need rights and rewards.

Where the need arises, they will perambulate, and perhaps run faster than we can.

Once we learn how to make self-replicating machines, we will no longer need to make anything ourselves. A single such machine might be despatched to construct an entire factory, given the resources and sufficient time. This is, after all, the way *life* works. It might take us time to do it as well.

The problem that remains, once you can automate all production, is how to distribute goods and services fairly. Owners of the factories will not profit if they alone retain their produce because their market will vanish. They may also face an angry mob at their door, demanding a share. Communism (where the State owns all production) has not been successful in the past, and led to totalitarianism (the opposite of democracy). This author, for one, has no wish to suffer dictatorship. Perhaps a system driven by the service ethic might work, where people volunteer to set up and/or run facilities for a year or so at a time, in return for no reward beyond their sense of usefulness and accomplishment.

Paradise might otherwise become very boring indeed.

Information technology is insufficient in itself. The chapter on space alluded to other knowledge vital to securing our future liberty and prosperity. To recap: to go to the Moon, we needed to recycle air; for an extended stay in orbit, aboard a "space station", we had to learn how to recycle water (as it's too heavy to resupply from the Earth below); to fly to Mars, we must learn how to recycle food *and* how to shield ourselves from strong radiation. The consequences of success here are three-fold.

Once we have learned how to do those things, we can live much more cheaply on Earth. Water and waste treatment and distribution, not to mention food production, are hideously expensive. The costs vary according to where you live. Recycling should eventually be far cheaper, and would allow us to live wherever we please. Pretty much the same is true of renewable energy production.

Solar panels began with the space programme.

The second consequence is that we need no longer rape our planet to survive and grow. The only limitation on population would be the availability of desirable places to live. But we already know that people tend to have fewer children (and better timed) the more prosperous and educated they are.

So there should be no need to fear over-population, any more than famine.

The last consequence is one we have mentioned before. Our descendents will wonder aghast at the dangers we faced, living out our days on the surface of a planet. Earthquake and tempest will pose no threat to them in their orbital habitats and space arks, enjoying the infinite resources out there.

Though war may remain with us till the end.

Application

4.15 For a scalable computing surface, we need a graph whose diameter (largest distance between two nodes) grows slowly with size (the number of nodes), with constant degree (number of arcs connecting each node).

What alternatives to a simple torus are there?

4.16 Suppose it does indeed become possible to automate anything, even the construction of a factory, while society finally succeeds in blessing everyone with both health and a three-tier education. Peace and prosperity emerge.

What would that society look like? What might it set out to collectively achieve, and why? What would remain to motivate individuals, and to what personal ends?

Index

www.ingramcontent.com/pod-product-compliance
Ingram Content Group UK Ltd.
Pitfield, Milton Keynes, MK11 3LW, UK
UKHW061952290726
14090UKWH00021B/1190